A Watershed in Global Governance?

An Independent Assessment of the World Commission on Dams

A Watershed in Global Governance?

An Independent Assessment of the World Commission on Dams

Navroz K. Dubash
Mairi Dupar
Smitu Kothari
Tundu Lissu

CAROL ROSEN
PUBLICATIONS DIRECTOR

HYACINTH BILLINGS
PRODUCTION MANAGER

MAGGIE POWELL
DESIGN AND LAYOUT

CAROLLYNE HUTTER
EDITOR

ISBN: 1-56973-494-1

Printed in India.

Production coordinated by Bindia Thapar and Navin K. Joneja

Cover Photo Credits:

Left: Commissioners at the South Asian regional consultation, Colombo, December 1999. Courtesy of the World Commission on Dams.

Center: Protest. Courtesy of Samfoto.

Right: WCD meeting in Tucurui, Brazil. Courtesy of the World Commission on Dams.

Contents

Acknowledgements

The authors are extremely grateful to the Commissioners, Secretariat staff, and Forum members for the time and assistance they generously provided in the course of this assessment. We also wish to thank our reviewers for their time and efforts. While they bear no responsibility for the final product, their comments have led to substantial improvements in the document. Our formal reviewers were: Hope Chigudu, The Global Fund for Women; Ken Conca, University of Maryland; Anthony Dorcey, University of British Columbia; M. Gopalakrishnan, Government of India; Shalmali Guttal, Focus on the Global South; Minu Hemmati, consultant to the UNED Forum; Andres Liebenthal, World Bank Operations Evaluation Department; Patrick McCully, International Rivers Network; Andreas Seiter, Novartis Corporation; Chaiyuth Sukhrsi, Mekong River Commission; and Richard Taylor, International Hydropower Association. Joji Cariño and Thayer Scudder, former Commissioners of the WCD, and Achim Steiner, Jeremy Bird and an anonymous reviewer of the WCD Secretariat provided informal review comments.

We are also grateful to several others who contributed knowledge and insights at key moments. Manuel Pulgar-Vidal of the Peruvian Society for Environmental Law attended and provided feedback on the WCD regional consultation in São Paolo in August 1999. Flávia Braga Vieira of the Federal University of Rio de Janeiro prepared a paper on the interaction of Brazil's Movement of Dam-Affected People (MAB) with the WCD, during a two-month research fellowship at WRI.

WRI thanks several colleagues, within WRI and across the world, whose contributions have enriched this report. This report would not have been completed without Fritz Kahrl, who co-ordinated research and production, provided editing and research support, and maintained enthusiasm for the process throughout. Frances Seymour, WRI's representative on the WCD Forum, encouraged us throughout this process, and was instrumental in helping us clear bureaucratic, financial, and other obstacles. She challenged us to strive for high standards of research and analysis, even while allowing us full independence to reach our own conclusions. Elena Petkova contributed ongoing advice, detailed review, and attended key WCD meetings on behalf of the assessment team. Thanks also to Nathan Badenoch, Don Doering, and Janet Ranganathan, who took time out of their busy schedules at short notice to provide helpful comments as part of WRI's internal review. Tony Janetos and Grace Bermudez steered the document through the review process. Danilo Pelletiere of George Mason University and George Faraday contributed greatly to the assessment by reviewing the global governance literature. Two interns performed well above and beyond the call of duty. Ray Wan of the Yale School of Forestry undertook extensive research on the WCD's media strategy and coverage of the WCD in the popular media, and Luna Ranjit, a recent graduate of Grinnell College, provided research assistance on the WCD's thematic reviews. To all these, and other colleagues who helped through encouragement, support, and filling in the gaps in our workload, we are most grateful.

Lokayan wishes to acknowledge several contributors and colleagues. Anil Bhattarai, a Nepalese action researcher spending a year at Lokayan, and Gopal Siwakoti "Chintan," director of INHURED International, Kathmandu, joined the Lokayan team at the inception of the assessment. Anil withdrew six months later to pursue advanced studies. He was replaced by Ramananda Wangkheirakpam, Jawaharlal Nehru University, whose diligence and consistency contributed significantly to our report. Lakshmi Rao, Jawaharlal Nehru University, joined the team for three months, primarily to assess the entire process of a WCD thematic study and help organise the India workshop on the WCD report. Biplove Chaudhary also contributed to the organising of this workshop. A special word of gratitude to Minar Pimple and Dilip Bhadarge of YUVA, Mumbai for their trust and support. Finally, without the sustained administrative and logistical support provided by P.T. George, the India process and the publishing of the final report would have definitely suffered. We are immensely grateful to all of them as well as to the other colleagues, Steering Committee members, and staff of Lokayan who patiently endured the onslaught in their midst.

LEAT acknowledges the invaluable contributions from the following: Melchisedeck Lutema, a LEAT staff attorney who was part of the original study team until July 2000 when he moved to the US for his graduate studies. Lutema rejoined the team as an intern from American University in Washington DC between January and April 2001. Susan Mlangwa-Nangwala, a sociologist from Uganda, prepared the Ugandan case study; Josephat Ayamunda, a Kenyan lawyer, did the Kenyan study, and John Langas Mughwai and Marcel Sawa Muro undertook the Tanzanian case study. Various LEAT colleagues gave ideas and comments at various stages of the study and to them all we are grateful.

All the authors would like to acknowledge and thank their families and loved ones who now, involuntarily, know more about the WCD than anyone could reasonably choose to. Mairi Dupar and Smitu Kothari would particularly like to thank their daughters, Heather and Emma. By accepting long workdays, enduring long absences, and tolerating distracted parents, our families supported us with great, and sometimes less, amounts of patience. It is now time to pay back the debt, a task we all commit ourselves to in the coming months.

We would like to thank Lokayan for its voluntary institutional partnership. We would also like to thank the New York and New Delhi offices of the Ford Foundation for providing major funding for the research. We would particularly like to thank Srilatha Batliwala, Lisa Jordan and Ujjwal Pradhan for their enthusiasm and support. Youth Unity for Voluntary Action (YUVA) received and coordinated the grant for the work in South Asia. WRI is also grateful for funds it received from the Royal Dutch Ministry of Foreign Affairs, the Swedish International Development Cooperation Agency, the U.S. Agency for International Development, and the MacArthur Foundation that were used in the start-up phase of this assessment. We would like to thank the WCD for providing funds for translation and dissemination of the assessment's findings, separate from research and preparation of the report.

N.D.
M.D.
S.K.
T.L.

Chapter 1

Introduction

An Experiment in Global Public Policymaking

In mid-2000, Medha Patkar, a leader of one of the best-known social movements in India, and Göran Lindahl, the Chief Executive Officer of one of the world's largest engineering firms, participated in a meeting together in Cape Town. The two came from different worlds. Ms. Patkar was weak from undertaking a hunger strike to protest a dam on the Narmada River in western India. Mr. Lindahl arrived at the last minute on his private jet. Before the meeting, Ms. Patkar animatedly described the recent protests, showed Mr. Lindahl pictures of the villagers, and narrated their experiences.

So began a typical meeting of the World Commission on Dams (WCD). Ms. Patkar, Mr. Lindahl, and their 10 colleagues from government ministries, the private sector, and civil society were all Commissioners on the WCD. Their common task was to address the conflicting viewpoints that have made large dams a flashpoint in the arena of environment, development, and justice.

The WCD was formed following a meeting of diverse dam-related stakeholders in early 1997 to discuss the past and future of large dams. The World Bank and the World Conservation Union (IUCN) initiated the process in response to growing protests at dam sites around the world. Although originally focussing on a study of the World Bank's dam-building record, the process grew into an independent review that consumed the time of 12 Commissioners, a full-time professional Secretariat, a 68-member advisory Forum, and thousands of contributors. The WCD's goals were to build a comprehensive knowledge base of large dams' development effectiveness and to develop criteria and guidelines to advise future decision-making on dams. *(See Box 1.1.)*

Box 1.1

Key objectives of the WCD

- A global review of the development effectiveness of large dams and assessments of alternatives.
- A framework for options assessment and decision-making processes for water resource and energy services and development.
- Internationally acceptable criteria and guidelines for planning, designing, construction, operation, monitoring, and decommissioning of dams.

Source: World Commission on Dams, Interim Report, July 1999.

The WCD was an extraordinary process in several regards. The Commission included voices that had previously been excluded from global commissions. It demonstrated both the feasibility and the challenges of consulting widely with the public and striving for transparency in a work programme. In spite of the challenges of broad representation, the Commission managed to produce a consensus report, *Dams and Development (see Box 1.2)*, which held considerable legitimacy because it was the joint work of Commissioners from diverse backgrounds.

Indeed, because of its efforts at representing a range of views, its emphasis on broad consultation, and its commitment to transparency, the WCD described itself as, and was proclaimed by others to be, a unique experiment in global public policymaking.[1] During the life of the Commission, multilateral institutions, governments, non-governmental organisations (NGOs), and corporations debated whether it was a model for global public policymaking in other arenas. Since then, discussions about the replicability of the WCD have cascaded into areas as diverse as extractive industries, trade and environment, food security and genetically modified organisms, and debt relief.

The Commission demonstrated the feasibility and challenges of broad consultation and transparency.

The final chapter on the WCD—concerning its impact on dam-related planning and practice and, therefore, its long-term effectiveness—will be written long after the report's release. Thousands of dam-related stakeholders around the world are already reaching for the Commission's findings, poring over them, and debating them publicly. The process outlined above was sufficiently robust that it engaged a variety of governments, international agencies, NGOs, people's movements, and private firms. Widespread engagement during the process led these diverse groups to take the report seriously and to recognise that the Commission spoke with a certain moral authority, even if its recommendations were not binding in the legal sense.

A Watershed in Global Governance? tells the story of the WCD experiment and assesses its implications for future global public policymaking. We examine how the WCD came about and how the commitment to good governance was infused in its work. We look at the practical challenges of implementing independence, transparency, and inclusiveness, and how the experience affected the WCD's legitimacy with stakeholders. Finally, we consider the different strategies for influence available to an advisory commission, such as the WCD, that has no binding mechanisms for compliance. We seek to locate the diverse reactions to the final report in the context of evolving norms of development practice.

The World Commission on Dams in Historical Context

The WCD emerged from several strands in the recent history of global policymaking. First, the WCD built upon a history of global commissions that have sought either to reconcile economic growth and environmental sustainability (such as the Brundtland Commission and the Stockholm and Rio Conferences) or to address North-South inequalities and questions of justice (such as the Brandt and South Commissions). Indeed, the WCD marked a step forward by incorporating at once the themes of social justice, human rights, ecological sustainability, and development in its work.

Second, the dams arena illustrates the growing ability of transnational civil society networks to contribute to global public policy agendas. The WCD was formed as a result of national and international civil society protest against large dams, which was often directed at such multilateral agencies as the World Bank.[2] The high transaction costs created by civil society dissent persuaded the World Bank and selected allies in international finance and industry that a new approach was required to move the dams debate forward.

Third, the WCD stood out from previous commissions in its diversity—including pro-dam lobbyists and anti-dam protesters—rather than limiting itself to participants from a broad middle ground. By the standards of global commissions generally, it also marked a notable departure from the "eminent persons" model of distinguished public servants. It comprised, instead, active practitioners whose personal legitimacy derived from their prominence in international stakeholder networks.

Fourth, the WCD was one of many government, private sector, and civil society dialogues on development policy that have proliferated since the landmark UN Conference on Environment and Development in Rio de Janeiro in 1992. By including a broad range of stakeholders, the WCD was a leading example of a "multi-stakeholder process." By including multiple perspectives, integrating diverse viewpoints early in a policy process, and building constituencies for implementation, multi-stakeholder processes are intended to provide a more inclusive and pragmatic form of policy formulation.[3] Some consultative processes involving civil society, business, and governmental actors have a direct input into policymaking.[4] However, many multi-stakeholder processes lack formal authority for decision-making and result in declarations, policy recommendations, and codes of conduct that are not legally binding. The WCD report joined a recent profusion of normative instruments and processes in international development that have no legal stature in themselves but are intended to be considered by legislators and to influence development practice.[5]

Finally, the WCD's structure and functioning responded to a broader call by civil society for transparency and inclusiveness in global governance. Before and since the WCD's formation, numerous protests and advocacy efforts by NGOs and social movements have sought to open up global decision-making about trade and investment rules, and associated labour, human rights, and environmental standards—decisions that are made behind closed doors and in the hands of the few, but affect the lives of millions. As a multi-stakeholder process whose objective was to address the source of past conflicts, the WCD committed explicitly to being transparent and open in its work.

The WCD stood out from previous commissions in the diversity of its composition.

The debate over large dams was ripe for the WCD's approach. Dams issues provide a microcosm of the changing political roles of the state, civil society, and the private sector in the rush toward a globalised world. Private financing is playing an increasing role, expanding the number of actors who hold leverage in dams planning and decision-making. Decisions about dams often involve governments, private firms, and international financiers—including bilateral aid agencies, multilateral development banks, export credit agencies, and commercial banks. Social movements and NGOs have criticised these actors' lack of transparency and have vocally resisted their decisions. The increase in number and scope of physical protests has brought added urgency to the dams debate. At the same time, the number of dams under planning and construction has rapidly fallen, as cost-effective alternatives to large dams have become increasingly available, especially in providing energy services. The controversy generated by large dams and the changing face of the dams industry provided compelling reason for both supporters and opponents of large dams, although wary, to come to the table. This is the context in which discussion began over the formation of an independent commission to address the dams debate.

Analytical Framework, Methods, and Outline

Analytical Framework

The ability to convene diverse actors and keep them constructively engaged is a core principle of multi-stakeholder processes such as the WCD. For such processes to be successful, stakeholders must feel that they have access to the process, that their voices are fully heard, and that their participation in the deliberations is meaningful. The potential benefits of these conditions are twofold: first, such processes are better informed, integrate diverse subjective viewpoints, and result in better outcomes. Second, inclusion builds constituencies for implementation.

In this report, we look at the efforts of the WCD and its initiators to create political space for diverse access to the process through

- full representation of relevant stakeholder groups on the Commission,
- independence from external influence,
- transparency to ensure the Commission's accountability to stakeholders' concerns, and
- inclusiveness of a range of views in compiling the knowledge base.

We assess how the WCD put these principles into practice and the effect of this experience on stakeholder perceptions of the WCD's legitimacy as the process unfolded. This approach was made possible by the time frame of our assessment, which was concurrent with the WCD.

We pay close attention to the political and practical trade-offs that the WCD faced in its efforts to create a representative, independent, transparent, and inclusive process. Because the WCD brought

Box 1.2

The WCD's findings and recommendations

Key message of the WCD

- "Dams have made an important and significant contribution to human development, and the benefits derived from them have been considerable.
- In too many cases an unacceptable and often unnecessary price has been paid to secure those benefits, especially in social and environmental terms, by people displaced, by communities downstream, by taxpayers and by the natural environment.
- Lack of equity in the distribution of benefits has called into question the value of many dams in meeting water and energy development needs when compared with the alternatives.
- By bringing to the table all those whose rights are involved and who bear the risks associated with different options for water and energy resources development, the conditions for a positive resolution of competing interests and conflicts are created.
- Negotiating outcomes will greatly improve the development effectiveness of water and energy projects by eliminating unfavourable projects at an early stage, and by offering as a choice only those options that key stakeholders agree represent the best ones to meet the needs in question." (p. xxviii)

Findings of the WCD

- "Large dams display a high degree of variability in delivering predicted water and electricity services—and related social benefits—with a considerable portion falling short of physical and economic targets, while others continue generating benefits after 30 to 40 years.
- Large dams have demonstrated a marked tendency towards schedule delays and significant cost overruns.
- Large dams designed to deliver irrigation services have typically fallen short of physical targets, did not recover their costs and have been less profitable in economic terms than expected.
- Large hydropower dams tend to perform closer to, but still below, targets for power generation, generally meet their financial targets but demonstrate variable economic performance relative to targets, with a number of notable under- and over-performers.
- Large dams generally have a range of extensive impacts on rivers, watersheds and aquatic ecosystems—these impacts are more negative than positive and, in many cases, have led to irreversible loss of species and ecosystems.
- Efforts to date to counter the ecosystem impacts of large dams have met with limited success owing to the lack of attention to anticipating and avoiding impacts, the poor quality and uncertainty of predictions, the difficulty of coping with all impacts, and the only partial implementation and success of mitigation measures.
- Pervasive and systematic failure to assess the range of potential negative impacts and implement adequate mitigation, resettlement and development programmes for the displaced, and the failure to account for the consequences of large dams for downstream livelihoods have led to the impoverishment and

together opponents in the dams debate as well as a broad political middle, including one group or perspective risked alienating another. In addition, the work of a commission is inevitably shaped by practical trade-offs. Funds, time, and the patience and perseverance of commissioners, staff, and stakeholders are real constraints on any such process, no matter how high the aspirations to good governance. The real measure of the WCD's success was whether it managed these trade-offs well enough to create space for engagement by a range of stakeholders that was sufficiently broad to promote its results.

Representation and good process are ultimately only means to influence policy and practice. Impact can be difficult to measure, because multi-stakeholder processes often do not have formal authority as decision-making bodies, but seek to shape outcomes through their influence as an advisory voice. In this study, we deploy multiple criteria for assessment of the Commission's likely impact. First, we examine whether and how the Commission achieved consensus. Without consensus, a commission will be seen to have reproduced divisions among stakeholders, rather than transcending them. Second, we ask whether and how the narrow consensus among the Commissioners

Box 1.2 continued

suffering of millions, giving rise to growing opposition to dams by affected communities worldwide.
- Since the environmental and social costs of large dams have been poorly accounted for in economic terms, the true profitability of these schemes remains elusive."(p. xxxi)
- "Global estimates of the magnitude of impacts include some 40-80 million people displaced by dams while 60 percent of the world's rivers have been affected by dams and diversions." (p. xxx)

Recommendations of the WCD

- "Clarifying the rights context for a proposed project is an essential step in identifying those legitimate claims and entitlements that might be affected by the proposed project—or indeed its alternatives." (p. 207)
- "Those whose rights are most affected, or whose entitlements are most threatened, have the greatest stake in the decisions that are taken. The same applies to risk: those groups facing the greatest risk from the development have the greatest stake in the decisions and, therefore, must have a corresponding place at the negotiating table." (p. 209)
- "Effective implementation of free, prior and informed consent marks a significant step forward in recognising the rights of indigenous and tribal peoples." (p. 219)
- "An early focus on options assessment will exclude most questionable projects. Those that emerge will enjoy wider public support and legitimacy. It can reduce delays and additional costs and conflicts, benefiting all those affected by a project." (p. 222)
- "A range of measures is available to enhance and restore ecosystems from their man-modified state, and many are already in use worldwide. Locally driven processes to establish the objectives of environmental flows will lead to improved and sustainable outcomes for rivers, ecosystems and the riverine communities that depend on them." (pp. 231, 239)
- "Regaining lost livelihood requires adequate lead time and preparation and therefore people must be fully compensated before relocation from their land, house or livelihood base. An overarching Compliance Plan is the best way to ensure that compliance activities and measures are effectively pursued and implemented, and should be developed for each project." (pp. 242, 247)
- "In many…cases retrofitting existing dams with more efficient, modern equipment and control systems has achieved significant improvements in benefits, extending facilities and optimising operations. While new supply options may be needed in many countries, restoring or extending the life of existing dams and, where feasible, expanding and improving services from existing dams provide major opportunities to address development needs." (pp. 226-7)

Excerpted directly from World Commission on Dams, Dams and Development: A New Framework for Decision-Making (London: Earthscan, 2000).

can eventually be translated into a broader consensus among stakeholders. In particular, we explore the role of good process in constructing broader stakeholder buy-in to a commission's recommendations and reflect on the implications of stakeholder support for adoption of recommendations.

Finally, this assessment looks at historical precedent through a detailed survey of past commissions, civil society advocacy efforts, global conferences, and multi-stakeholder processes. All of these arenas represent important influences in the formation of the WCD. This rich past record provides a useful context for the assessment as it reflects the many strands that shaped the WCD. It provides a lens on the practical feasibility of different forms of stakeholder consultation and representation by demonstrating what has been accomplished before.

Approach and Methods

Three non-governmental organisations—the World Resources Institute (United States), Lawyers' Environmental Action Team (Tanzania), and Lokayan (India)—undertook this assessment. Our assessment team was structured to incorporate a diversity of Northern and Southern perspectives.

Within each organisation, there were one or two primary authors, supported by research assistants and field-based research fellows from Brazil, Kenya, Nepal, Peru, Tanzania, and Uganda.

The ability to engage diverse actors constructively is a core principle of multi-stakeholder processes.

The assessment maintained editorial independence from the WCD, although we had the WCD's co-operation. The work was funded almost entirely from private foundation sources, which are detailed in the acknowledgements section of this report.

The World Resources Institute was a member of the WCD Forum, in which it was classified as a "Research Institute." The Institute's participation in the Forum was kept separate, in terms of personnel and funding, from the work of this assessment in order to maintain an independent perspective on the WCD process.

Research for the assessment comprised semi-structured interviews, observation at WCD meetings, and analysis of WCD documents. We relied on interviews to reconstruct events, to capture varying perspectives on WCD proceedings, and to solicit reactions to the process and the final report. Over a two-year period, the research team interviewed 10 Commissioners, all Senior Advisors at the Secretariat, representatives of almost all 68 member organisations of the Forum, and several consultants and financial donors to the WCD. We also sought the views of dam-related stakeholders who were not formally involved in the WCD process. For example, we interviewed displaced people in Egypt and India to understand their perspectives on large dams and determine the accessibility and relevance of the WCD to them. Following the launch of the WCD report, we interviewed stakeholders from government and multilateral agencies, industry trade groups, and civil society organisations to document responses to the recommendations. In India, the team organised a large multi-stakeholder consultation for feedback on the report that included members of the Secretariat and former Commission. In East Africa, the team conducted small focus groups and interviews with stakeholders in Kenya, Tanzania, and Uganda.

Because participants in the process were highly sensitive to the WCD's charged politics, we decided to conduct interviews on a not-for-attribution basis to encourage candour from interviewees. To ensure accuracy, information gathered through interviews was checked against documentation where possible and against information provided in other interviews. In addition, we were attentive to the context and stakeholder location of the interviewee. Summaries of interviews, including relevant quotes, were circulated among team members in order to cross-check information.

The assessment team was allowed to participate in WCD Forum meetings, case study meetings, and regional consultations as observers. Appendix 5 lists meetings attended by members of the team. Although we requested permission to attend Commissioner meetings, the Commissioners declined this request in order to maintain the confidentiality of their deliberations. We were, however, granted access to the Secretariat's minutes of the Commissioners' meetings on the basis that the minutes be used as background information and not for citation.[6] During the meetings, in addition to conducting individual interviews, we noted how the meeting was structured, the information available, the role of the Commissioners, Secretariat, Forum members, and participants, the issues deliberated, and the tone and content of debates. This information provided valuable content, particularly for our assessment of inclusiveness and transparency. To complement our scrutiny of the WCD process from the inside, we analysed media coverage of the WCD.

To assess the voluminous work programme, we studied a sample of the Commission's case studies and thematic reviews. In examining the thematic studies, we drew on documentary records, including reviewers' comments, to assess how the scope of the study was shaped. We also interviewed Secretariat staff, consultants, and reviewers about their role in the thematic review process.

In early 2001, toward the end of our research process, we presented our preliminary findings to the WCD Forum members at their final meeting in

Spier Estates, outside Cape Town, South Africa. Their questions and comments were important in shaping our draft report. We subsequently posted the preliminary findings on our website, www.wcdassessment.org, and invited comments from concerned readers. We met three times as a full assessment team, and on several other occasions at WCD meetings to analyse the findings and draft chapters.

In the course of conducting the research, we refined and adapted our questions to react to information we received over this period. Our original research framework aimed to assess the effect of the WCD's work on strengthening or undermining emerging norms of global governance—such as the principles of openness, participation, and transparency to which the WCD was committed. Over time, we came to appreciate the importance of the composition of the Commission, Secretariat, and Forum to perceptions of the Commission's legitimacy. In addition, although we focussed on the process by which the WCD conducted its work, we also placed additional emphasis on the links between process and stakeholder reactions to the final report. We looked at the extent to which these stakeholder reactions invoked the WCD's process, to help test our hypothesis that good process builds constituencies for implementation.

We submitted drafts of this report to two rounds of peer review. Colleagues in each of our organisations conducted the first round of review. We then submitted the manuscript to 10 external reviewers. These included five persons with central roles in the WCD process, and another five with experience in global governance processes, or who were working in issue areas at the intersection of environment, development, and justice. In addition, we submitted the draft informally to five reviewers drawn from the Commission and Secretariat. These comments are available on the assessment team's website. Based upon these comments, we revised the report again before final publication.

As an effort to assess the WCD as a model for global governance, this study has potential implications for the WCD's broader legacy. This legacy is of considerable interest to stakeholder groups whose interests are affected by the Commission's perceived legitimacy. As a result, this study has come under the same kind of scrutiny and lobbying pressure by different interest groups as did the WCD's process itself, albeit not with the same intensity. We have been well aware of this dynamic during our research and the threats to independence that close engagement with stakeholders can bring. However, we are convinced that immersion in the process brings richness to the analysis, and when combined with awareness and critical self-reflection, the benefits of this approach outweigh the costs.

Outline

This report adopts a broadly chronological organisation of the formation and history of the WCD. Chapter 2 locates the WCD in the evolution of global governance efforts, and in multi-stakeholder processes in particular. Chapter 3 examines the WCD's origins to chart the unique aspects of the Commission's formation, including the reasons why different stakeholders in the dams debate were ready to sit down at the table together.

Chapters 4 to 7 are dedicated to an analysis of the Commission's structure and functioning. Chapter 4 examines how the WCD's creators tried to represent the full range of interests in the dams debate on the Commission, Secretariat, and Forum. Chapters 5 and 6 explore how the WCD's organisation and work programme were designed to promote inclusiveness, transparency, and independence and how these designs worked out in practice. Chapter 7 reconstructs dynamics within the Commission, and among the various organs of the Commission in exploring the trajectory toward a consensus. Chapter 8 reviews stakeholder reactions to the WCD final report and highlights factors that contribute to stakeholder adoption of the Commission's recommendations.

In Chapter 9, we return to the analytical framework described above. We assess the degree to which the WCD's track record in good governance contributed to stakeholders' willingness to engage with and act upon the Commission's final report.

Endnotes

1. See, for example, WCD Newsletter No. 3, June 1999. Online at: www.dams.org/newsletters/newsletter3.htm (23 August 2001). External audiences have echoed this framing of the WCD. See Jörg Baur and Jochen Rudolph, "A Breakthrough in the Evolution of Large Dams? Back to the Negotiating Table," *D+C Development Cooperation*, No. 2, March/April 2001, pp. 9-12. Online at: www.dse.de/zeitschr/de201-3.htm (28 September 2001); David Seckler and Achim Steiner, "More Crop per Drop and Dams on Demand? Implications for the 21st Century." Report given at the ODI-SOAS Meeting Series, 9 February 2000. Online at: www.oneworld.org/odi/speeches/water3.html (28 September 2001).
2. See Jonathan Fox and L. David Brown, eds. *The Struggle for Accountability: The World Bank, NGOs and Grassroots Movements* (Boston: MIT Press, 1998); Robert Wade, "Greening the Bank: The Struggle over the Environment, 1970-1995," in *The World Bank: Its First Half-Century.* Devesh Kapur, John P. Lewis, and Richard Webb, eds. (Washington: Brookings Institution, 1997).
3. Minu Hemmati et al., *Multi-Stakeholder Processes for Governance and Sustainability Beyond Deadlock and Conflict* (London: Earthscan, 2001). Online at: www.earthsummit2002.org/msp/ (28 September 2001).
4. For instance, at the meetings of the United Nations Commission on Sustainable Development (CSD) the results of the multi-stakeholder dialogues at the beginning of the sessions are summarised by the CSD Chairperson. These summaries are presented to negotiators the following week and assume the status of an official document. The delegates choose paragraphs from the summaries in formulating the formal decision. Personal communication with UNED Forum, 30 July 2001.
5. See A. Florini, ed. *The Third Force: The Rise of Transnational Civil Society* (Washington, DC: Carnegie Endowment for International Peace, 2000); Dinah Shelton, ed. *Commitment and Compliance: The Role of Non-binding Norms in the International Legal System* (Oxford: Oxford University Press, 2000).
6. In a few limited cases, sections of the minutes had been obscured by the Secretariat to preserve confidentiality.

Chapter 2

Multilateral Processes, Global Commissions, and Global Governance

For much of the past century, the history of global governance has been the history of intergovernmental processes. These processes have been organised around exclusive deliberations of officials and politicians with similar backgrounds—whether initiated by clusters of governments, intergovernmental agencies, or distinguished individuals. The outcomes were targeted at governments and international agencies, which were considered the only legitimate actors on the global stage.

At the same time, over the past two decades there has been an unprecedented growth of actors in civil society extending and building alliances and coalitions that transcend national boundaries. In the countries of the less industrialised world, civil society mobilisation grew as the post-colonial nation-building euphoria gave way to disillusionment with the capacity of state regimes to vigorously pursue policies of social transformation in favour of historically less privileged and marginalised peoples. Some of this mobilisation extended across national boundaries in response to adverse social, economic, and ecological impacts of the policies of multilateral financial institutions and the economic, political, and military activities of governments.

The emergence of civil society at the global level has both been facilitated by the telecommunications revolution that accompanied globalisation, and been spurred by the challenge of effective governance in a globalising world.[1] These alliances and coalitions have spawned on a diverse bed of issues—from land mines to nuclear weapons, from the global trade in animal skins to the control and ownership of genetic resources. Of particular relevance to this assessment, civil society alliances have also formed around development projects and the accountability of international institutions. Often, debate over these issues has led to the creation of regional or global fora, which have taken the form of commissions, tribunals, and working groups..

In reaction to the increasing number of actors active in shaping public policymaking, national governmental agencies and intergovernmental bodies have become more open to including non-state actors in a more structured manner. A popular approach has been to establish "multi-stakeholder processes," or "MSPs," that provide space for dialogue among a range of actors from various sectors of society, as part of a decision-making or advisory process.

Over time, the international system has been compelled to recognise emergent voices from civil society.

The formation of the World Commission on Dams (WCD) draws from a history of global commissions in the last quarter of the 20th century, from the expanding role of civil society actors as agents of change at the global level, and from the emergence of MSPs. In this sense, it is a step forward in the history of multilateral policymaking. In this chapter, we briefly review the contemporary history of global governance to illustrate both the exclusiveness of state-led international governance and the inroads made by civil society. Next, we trace the impact of transnational civil society organisations on setting agendas at the international level. We then discuss emergent forms of governance, MSPs, and the implications of these forms for effective compliance with norms of international governance. Finally, we locate the WCD process within these diverse strands.

From State-Centred Multilateralism to Multi-stakeholder Processes

Eminent figures dominate the history of global governance through the second half of the 20th century. In most cases, these people drew their credibility from positions of political power within national governments and international bureaucracies or their national or global moral stature. Based on their credibility and often backed by the authority of intergovernmental bodies, statesmen and stateswomen established a series of commissions to deliberate on the weighty issues of the day.

In this section, we describe the gradual opening of global discussions to embrace the views and participation of a broader set of actors. In telling this story, we also trace the shifting scope of concerns around global governance. In particular, we trace the emergence of concerns around environmental issues, and the progressive interweaving of environment with development and justice concerns.

Formation of the United Nations and Early Patterns of Public Participation

The architecture for current forms of global governance was established in the World War II years through a series of conferences and conversations among leaders of the Allies.[2] In the post-war period, the United Nations (UN), the International Bank for Reconstruction and Development (World Bank), and the International Monetary Fund (IMF) were established.[3] In addition, agreements and institutions in the areas of food, culture, and education were also discussed, leading to the creation of such influential agencies as the World Health Organization (WHO) and the United Nations Educational, Scientific, and Cultural Organization (UNESCO). This period also saw the the fledgling United Nations General Assembly formulate and adopt the Universal Declaration of Human Rights.

Although the UN was conceived as, and remains, predominantly a forum for inter-governmental engagement, it also provided space from its inception for consultation with non-state actors. Article 71 of the UN Charter allows the UN Economic and Social Council (ECOSOC) to establish "suitable arrangements for consultation," but only with "international" NGOs.[4] The ECOSOC created a three-part categorisation for NGOs in 1950, subsequently refined in 1968, based on the perceived degree of expertise of the NGO with the issues on the ECOSOC agenda. These rules applied progressively more stringent criterion for participation, from the "general" to the "special" to the "roster" category of ECOSOC consultative status,[5] and governed participation in UN meetings for the first half-century of the UN. It was only after the considerable NGO interest in, and engagement with, the series of UN-sponsored conferences in the 1990s that these rules were modified in 1996 to allow national NGOs to be eligible for consultative status with ECOSOC.[6]

Among post-World War II institutions, the International Union for the Conservation of Nature (IUCN—now called the World Conservation Union) was significant both for its focus on environmental concerns and for its broad membership. IUCN was established by Julian Huxley, the first director of UNESCO. A joint initiative of UNESCO, the government of France, and the Swiss League for Nature Protection, IUCN is one of the few international institutions that formally bring together stakeholders from within and outside government. From its inception when it started with 80 members, IUCN now brings together 78 states, 112 government agencies, 735 NGOs, 35 affiliates, and some 10,000 scientists and experts from 181 countries in a unique worldwide partnership. IUCN members rejected a proposal in 1994 to include a membership category for "industry groups."[7] IUCN's historical focus has been on conservation of species and ecosystems, which has recently expanded to include related human development issues. It has had considerable success in promoting conservation, spearheading several key international agreements in the 1970s and 1980s. Of central relevance to this study, IUCN was also one of the two convenors—the World Bank was the other—of the WCD process.

Currents of Change: Economic Injustice and the Limits and Use of Nature

The 1960s and early 1970s saw two separate currents of change that developed from national contexts to become global phenomena. Several countries had emerged from colonialism in the

Box 2.1

Chronology of major global commissions and conferences

1945	United Nations formed
1968-69	Commission on International Development (Pearson Commission)
1972	Stockholm Conference on the Human Environment
1977-79	Independent Commission on International Development Issues (Brandt Commission)
1980-82	Independent Commission on Security and Disarmament Issues (Palme Commission)
1983-87	World Commission on Environment and Development (Brundtland Commission)
1987-90	South Commission
1992	United Nations Conference on Environment and Development (Earth Summit)
1992-95	Commission on Global Governance
1995-98	World Commission on Forests and Sustainable Development
1995-98	Independent World Commission on Oceans
1998-2000	World Commission on Water

1950s and 1960s, with high aspirations for their people. Yet, many countries remained mired in poverty, spurring the UN to declare the 1960s a "Decade of Development." At the same time, Western Europe, the United States, and Japan were enjoying a period of unprecedented prosperity. The 1969 Pearson Commission, established by the World Bank to investigate Third World poverty, noted an "atmosphere of decreasing interest for development assistance" in the industrialised world and "signs of dejection and growing impatience" in the developing world. The Commission called for trade measures that favoured developing countries, promotion of foreign direct investment to those countries, and an increase in development assistance to 0.7 percent of donor country GNP.[8] Increasingly impatient, in 1974 the developing countries in the UN—the Group of 77 (G77)—called for a New International Economic Order (NIEO) based on a rise in raw material prices, debt reduction, and more favourable conditions for the transfer of technology.[9] Questions of economic justice and development were firmly on the agenda by the early 1970s.

In the 1960s, the ecological limits and costs of the singular pursuit of economic growth were becoming increasingly apparent in the North, while the South was beginning to recognise the integral role that the environment played in the subsistence of its peoples. The publication of *Silent Spring*[10] in 1962, which carefully and eloquently documented the effects of pesticide pollution, is considered a founding event of American environmentalism.[11] In both Germany and the UK, this period witnessed the formation of Green Parties.[12] The early 1970s also saw the publication of the Club of Rome's (COR)[13] path-breaking report, *Limits to Growth*,[14] and *The Ecologist*'s "Blueprint for Survival."[15] Based on computer simulations, these reports warned of dire effects from continued exponential growth in five interconnected trends of global concern—industrialisation, population growth, widespread malnutrition, depletion of renewable resources, and ecological damage.

Although the industrialised world focussed on concerns from pesticide pollution to the finiteness of natural resources, in the less industrialised world, there was growing concern over the inequity in access to and control over productive natural resources.[16] In a burgeoning "environmentalism of the poor", these concerns were perhaps best illustrated by the Chipko movement against logging in India and the rubber tappers' mobilisation in the Brazilian Amazon in the early to mid-1970s.[17]

Both economic and ecological concerns were brought into stark relief by the Arab nations' 1973 oil embargo. Within a year, the price of crude oil tripled, causing an increase in the prices of other raw materials and goods and services. The recession and inflation that followed fuelled widening recognition of the vulnerability of industrialised and less industrialised country economies both in their sources of energy and in the finiteness of their non-renewable resources. By demonstrating the destabilising effects of sharp shifts in natural resource prices and the importance of control over natural resources, the embargo also focussed and polarised debates over North-South economic justice. The stage was set for a series of charged global deliberations on the twin, but as yet rarely connected, issues of economic justice and environmental concerns.

Change from Above?

Based on its dismal findings, the Pearson Commission concluded with a call for a second, more effective UN-led "Decade of Development." In response, in the early 1970s, the UN organised a series of conferences: on environment in 1972; on population and food in 1974; and on women in 1975.[18]

The Stockholm Conference on the Human Environment (1972) marked the international coming of age of environmental concerns, but also was suffused with the North-South conflicts of the time.[19] The agenda primarily reflected Northern concerns of scarcity and pollution. Southern countries were deeply suspicious that the environmental agenda would force them to slow down their processes of industrialisation and economic development. In the memorable, if misguided, words of the Indian Prime Minister, Indira Gandhi, the Southern countries argued that "poverty is the greatest polluter." In echoes of the concurrent debates over the international economic system, they argued that the economic order did not grant them economic independence to complement the political independence won through hard-fought anti-colonial struggles. Despite these tensions, the Stockholm Conference not only placed an important set of issues at the intersection of environment, development, and justice on the international agenda, but also established the United Nations Environment Programme (UNEP) to carry the environmental agenda forward.

In addition, the Conference was notable for the high level of non-governmental participation and interest. A total of 255 NGOs were accredited to the Conference.[20] Many of these had no prior connection to the UN and used their access to provide immediate and often critical commentary on the process to the outside world, establishing a tradition that has been carried on by environmental organisations at international meetings ever since.[21] Yet, during this period, most NGO participants were from the North; Southern NGOs accounted for only 10 percent of the NGOs present.[22]

Eminent figures dominate global governance through the 20th century.

The tension between environment and development continued to occupy a prominent place on the global agenda, fuelled by the Stockholm Conference and promoted by the newly created UNEP. In 1974, UNEP and the UN Conference on Trade and Development (UNCTAD) jointly sponsored a symposium in Cocoyoc, Mexico on the "Pattern of Resource Use, Environment, and Development." The symposium's Declaration highlighted the need for greater self-reliance by poor countries and called for changes in the international order to enable this. The Declaration further stated that "Human beings have basic needs… Any process of growth that does not lead to their fulfilment—or even worse, disrupts them—is a travesty of the idea of development." [23] As with Stockholm, the environment and development debate was suffused with North-South concerns over economic justice.

Not all the international fora organised at this time were characterised by the same degree of openness as Stockholm. At the same time as United Nations organisations were initiating several processes, other more closed and elite-led efforts at global coordination were also evident. A brief detour through the formation of the Trilateral Commission indicates the tenor of parallel efforts at shaping global trends. The Trilateral Commission was the brainchild of David Rockefeller, who had in mind "a private organisation, whose primary objective…would be to bring the best brains in the world to bear on the problems of the future." This group of leading private citizens from Europe, the

United States, and Japan came together in 1973 to deliberate on leading issues of the day, including international trade and investment, environmental problems, crime and drugs, population control, and assistance to developing countries.[24] The influence of the Commission was based on its ability to assemble the "highest level unofficial group possible."[25] In this it was successful, with successive Commissions (which continue to be convened on a triennial basis) consisting of senior government officials and corporate leaders, all acting in their personal capacities. The model for change implicit in the Trilateral Commission is that it provides a vehicle to directly shape the opinions, and hence the actions, of those with political power.[26]

The Stockholm Conference on the Human Environment marked the international coming of age of environmental concerns but was suffused with North-South conflicts.

By contrast, the Independent Commission on International Development Issues, established in 1977 and headed by former German Chancellor Willy Brandt, was firmly rooted in the formal system of international governance embodied in the United Nations. At the same time, it conformed to past practice in its dependence on participation of eminent persons for both legitimacy and content. Although the Commission was established on the recommendation of World Bank President Robert McNamara, the Secretary-General of the UN expressed support for the work of the Commission. In terms of its scope, the Brandt Commission was the direct descendant of the Pearson Commission and the NIEO discussed at the UN in 1974. The Commission's mandate was "to study the grave global issues arising from the economic and social disparities of the world community" and "to suggest ways of promoting adequate solutions to the problems involved in development and in attacking absolute poverty."[27]

The Commission's 19 members were carefully balanced between North (9) and South (10) by the Chairperson, who was "anxious that the Third World members…not be in a minority position."[28] Its ranks were filled with dignitaries, including three former prime ministers, one former president, seven former ministers, and other ambassadors and senior members of national and international governmental bodies. In keeping with common practice for such commissions, each commissioner acted in his or her individual capacity. Aside from one Commissioner with media experience and a few with some background in the private sector, the Commission lacked representation from non-governmental actors. The work of the Brandt Commission was organised around a roster of eminent persons—statesmen or noted intellectuals—who were invited to testify or submit their views for consideration to the Commission. In addition, members travelled to capital cities to meet with presidents, prime ministers, and heads of regional and international organisations.

The Commission published two reports, *North–South: A Program for Survival* and *Common Crisis*, in which it made comprehensive proposals. The Commission's recommendations are startling in their scope. They include: a World Development Fund to which communist nations would have access; a tax on trade, minerals from the sea, and weapons sales, with proceeds going to poorer countries; an agreement on the production, pricing, supply, and conservation of oil; and a transfer of resources to the South by increasing development aid to 1 percent of GNP of donor countries by 2000 and by expanding the capital flowing through the World Bank and the IMF.[29] Despite the breadth and comprehensiveness of its exercise, most governments remained indifferent to the Commission's recommendations. Nonetheless, the Commission marked a change in the international community's response to global issues. It gave space to such ideas as human needs, self reliance, respect for local cultures, and the extension of participation and representation to communities. Ecological issues were only minimally acknowledged.

Perhaps the most significant legacy of the Brandt Commission is that its composition was based on one criterion of representation—equal representation of industrialised and less industrialised governments in global policymaking processes. By some reports, this resulted in a stormy internal North-South dialogue, which was resolved only when a spokesperson from each side assumed responsibility for negotiating a final report.[30]

By the 1980s, several UN agencies collaborated to bring governments on board and open policy processes to civil society and community organisations. In 1980, UNEP launched the World Conservation Strategy jointly with IUCN, the World Wildlife Fund, the Food and Agriculture Organisation of the United Nations (FAO) and UNESCO and developed a Global Framework for Environmental Education and a Global Environmental Monitoring System (GEMS). The preparation and publication of the World Conservation Strategy popularised the term "sustainable use."[31]

The Rio Summit was inclusive compared to past UN events.

After a two-year Independent Commission on Security and Disarmament led by Olaf Palme between 1980 and 1982,[32] the focus shifted squarely back to environmental concerns. In 1983, the UN General Assembly established the World Commission on Environment and Development (WCED) with former Norwegian Prime Minister Gro Harlem Brundtland as the Chairperson. The 23 Commissioners were drawn from senior levels of national government, international organisations, and academia. This time, there was no representation from the private sector or civil society. This absence was striking, since the mandate of the WCED included raising the level of understanding and commitment to action of various sectors of society, including business and voluntary organisations.[33]

However, the WCED distinguished itself from prior commissions with a commitment to an "open, visible, and participatory" process. It put this commitment into practice through a series of public hearings in eight countries involving consultations with hundreds of representatives from governments, scientific research institutes, companies, and NGOs, as well as with the general public.[34] The body of documents studied by the Commission—more than 10,000 pages—was brought together in a Collection of the Archives of Sustainable Development. Copies of this collection were placed in six academic centres throughout the world.[35]

The WCED's report, *Our Common Future*, legitimised the concept of sustainable development as "a form of development that meets the needs of the present without compromising the ability of future generations to meet their own needs," and tried to make the links between environment, development, and poverty.[36] The Commission forcefully acknowledged that the interface of environment and development needed the engagement of a wide cross-section of actors—individuals, NGOs, governments, and international organisations. However, the Commission was also criticised for failing to probe the processes that generated poverty in the first place, and for neglecting the fundamental relationship between social equity and sustainable development.[37]

Although the WCED perpetuated the model of global governance from above, dominated by eminent persons, it did acknowledge and put into practice a substantial consultative role for the broader community of stakeholders. In addition to having bequeathed the concept of sustainable development, the WCED's procedural legacy is its comprehensive system of public hearings.

The North-South tensions so apparent in the Brandt Commission and echoed in the Brundtland Commission continued to fester. These tensions included growing disparities among and within nations, the predominance of unequal models of development, and the increasing fragility of natural resources. In 1986, the Non-Aligned Movement announced the formation of a South Commission, which was established in 1987 under the chairpersonship of former Tanzanian President Julius Nyerere. This Commission, consisting exclusively of industrialising countries, started with the belief that eloquent agreements on the common heritage of humankind were not going to change a situation primarily caused by the powerful nations and vested interests. The Commission sought to make a "case for self-reliant, people-centred development strategies."[38]

The South Commission followed the established practice of selecting "distinguished individuals" as Commissioners.[39] However, among these were a sprinkling of individuals from churches, academia, and NGOs. The South Commission also followed the established route for its work programme, creating working groups on topics such as debt, the Uruguay Round of trade negotiations, and North-South relations; forming groups of outside experts to complement its studies; and holding meetings with officials and intellectuals in various regions. In its report, *The Challenge to the South*, released in 1990, the Commission established a vision for development as a "process that enables human beings to realise their potential, build self-confidence, and lead lives of dignity and fulfilment." To achieve this vision, the South Commission called for concerted national efforts to harness the potential of citizens, and for greater political and economic co-operation in the form of enlarged South-South co-operation.[40] In addition, the Commission noted that international arrangements for trade, finance, and technology could handicap the South and argued for a cogent Southern stance in North-South deliberations to reform these arrangements. The South Commission stood out from past commissions largely because it was an initiative of, and by, Southern nations. In its structure and functioning, apart from a somewhat greater representation of civil society, it followed the model of commissions before it.

It took a second major conference on environment and development to establish a significantly expanded space for non-state actors in global governance. In its final chapter, the Brundtland report called for an international conference to review progress in sustainable development and create a follow-up structure. Based on a formal resolution by the UN General Assembly, the United Nations Conference on Environment and Development (UNCED), or Earth Summit, was held in June 1992 in Rio de Janeiro, Brazil.

The Rio Summit was, compared to past UN events, the most inclusive ever. It created new legitimacy and political space for NGOs and peoples' representatives. More than 1,400 NGOs registered for the summit, with more than a third from the South, making it the largest ever face-to-face gathering of Northern and Southern NGOs.[41] The UN created an unprecedented accreditation process that gave NGOs significant access to the formal events.[42] Moreover, several governments had NGO representatives as part of their delegations, although some were sceptical of the "representativeness" of the individuals present.[43] The parallel Global Forum attracted 9,000 groups. Many participated in the International NGO Forum, which drafted 39 "Alternative Treaties" as exercises in direct citizens' diplomacy. The goal was to produce agreements on actions that citizens' groups themselves would undertake.[44] Those who could not physically participate contributed through Econet, a new electronic medium.[45] Arguably even more significant than participation by civil society and the private sector in the meeting was the role these non-governmental actors played in the two-year preparatory process and in shaping the agenda leading up to the Earth Summit.[46]

NGOs enjoyed only limited access to the formal deliberations. Moreover, because the main channel for influence was through participation in national delegations, relatively mainstream organisations that were more likely to be invited on to delegations were at a comparative advantage. More explicitly, political NGOs found fewer opportunities for expression in the technocratic nature of the proceedings, as did "consciousness-raising" NGOs who sought social change through education and empowerment.[47] NGOs mirrored North-South tensions that ran through official government positions.

At the Earth Summit, more than 100 heads of state met to address urgent problems of environmental protection and socio-economic development. The assembled leaders signed the Framework Convention on Climate Change and the Convention on Biological Diversity, endorsed the Rio Declaration and the Forest Principles, and adopted Agenda 21, a 300-page plan for achieving sustainable development in the 21st century. The Commission on Sustainable Development (CSD) was created after UNCED to monitor and report on implementation of these agreements.[48] In addition to these significant outcomes, the Earth Summit marked an important milestone in the inclusion of civil society voices in multilateral processes.

In 1992, there was an explicit attempt to pull together the strands of the various commissions of the previous 20 years in the form of the Commission on Global Governance. Initiated by Willy Brandt, the origins of this Commission were in a meeting attended by former commissioners from the Brandt, Brundtland, Palme, and South Commissions. The decision to set up a Commission on Global Governance, co-chaired by Ingvar Carlsson, former Prime Minister of Sweden, and Sridath Ramphal, former Secretary-General of the Commonwealth, was endorsed by the UN Secretary-General. As with all its predecessors, the Commission was composed of eminent individuals, with only 1 individual among the 26 whose background was not dominated by government service.[49] The scope of the Commission broadly embraced that of all the commissions before it and included security, democracy and the role of civil society, co-ordination in economic policy, poverty alleviation, and environment. The goal was to "develop a common vision of the way forward for the world in making the transition from the Cold War and in managing humanity's journey into the twenty-first century."[50]

In addition to organising working groups and requesting research institutes to organise seminars, the Commission held four briefing meetings for NGOs in Geneva. In addition, it set aside time for meetings with NGOs as part of Commission meetings held in New York, Mexico City, Tokyo, and Delhi. These minimal efforts did not signal a significant attempt at inclusion; the Commission on Global Governance remained a closed group of world leaders.

In its 1995 report, *Our Global Neighbourhood*, the Commissioners wrote that they were convinced that the world is ready to accept "a set of core values that can unite people of all cultural, political, religious, or philosophical backgrounds. It is fundamentally important that governance should be underpinned by democracy at all levels and ultimately by the rule of enforceable law". UN agencies, such as UNCTAD and the UN Industrial Development Organization (UNIDO), as well as civil society, criticised the report as being Northern in orientation and weak in its analysis of the social, economic, and ecological costs of the present process of economic development.

From 1995-2000, three sector-specific commissions—on forests, oceans, and water—were established and ran their course, in some cases explicitly following in the footsteps of past commissions. Of these, the World Commission on Forests and Sustainable Development located itself most directly in the tradition of past debates over environmental concerns. The Commission, established in 1993 out of debates at the Earth Summit,[51] focussed on the political rather than technical issues from those discussions. In addition, it held a series of five regional hearings modelled after the Brundtland Commission. Its 24 members were largely eminent persons. At the conclusion of its work in 1998, the Commission left a legacy of a four-part Forest Trust, composed of a Forest Watch, a Management Council, an Ombudsman, and a Forest Award.[52]

The Independent World Commission on Oceans, established in 1995 by Mario Soares, former President of Portugal, released its report in 1998, timed to coincide with the UN-declared International Year of the Ocean. The Commission's origins were firmly rooted in the UN system. The dauntingly large 43-member Commission included the familiar range of ministers, parliamentarians, and ambassadors, with a sprinkling of academics and technical specialists, and relied largely for its findings on study groups led by experts.[53]

The environment arena has increasingly been influenced by debates over development and justice.

The World Commission on Water (described in further detail in Chapter 6) was established in 1998 by the World Water Council, and ran concurrently with the WCD. Its 28 members included some NGOs and research organisations, in addition to professional associations, government representatives, and international organisations.[54] In this sense, the Commission's composition was arguably less centred on international eminence than past processes and more guided by practitioners, albeit with a heavy representation of international aid and water bureaucracies. The Commission was tied to a concurrent World Water Vision process aimed at developing a vision for addressing global water scarcity. The vision was developed through a multi-stakeholder consultative process organised around regions and sectors. NGOs critical of the process charged that this process tended toward expert participation with little inclusion of civil society, and that aid agencies controlled the effort.[55]

What emerges from this review of focal points of global governance during the last third of the century? First, the structure of commissions left little space for engagement with non-governmental actors. The space that did exist was far larger and more genuinely consultative in discussions around environment and development issues than around economic justice. By the late 1990s, the structure of regional hearings, first established by the Brundtland Commission, had become accepted practice, as illustrated by the commissions on forests and water. Second, although debates over international economic justice have remained unconcerned with environmental questions, the environment arena has increasingly been influenced by debates over development and economic justice. Third, the global fora described above—with the exception of the Trilateral Commission—were designed to affect the actions of world institutions and structures.

Yet, the impact of most global commissions has been slight. The Brandt and South Commissions came out with concrete recommendations that were not implemented. The Brundtland Commission succeeded in popularising the term "sustainable development" and raising popular awareness of issues at the environment and development interface, but resulted in few concrete changes. The Stockholm and Rio Conferences were more successful. The first resulted in a new and important UN agency, and the second served as a catalyst for a series of important global environmental negotiations, whose future is still not assured. The three recently concluded sectoral commissions, on forests, oceans, and water, have yet to demonstrate any lasting effects.

Finally, distinguished individuals have dominated global commissions, with eminence being almost synonymous with high-level experience in government and intergovernmental bodies. Indeed, the same individuals have served on many of these commissions, as a few examples illustrate. Willy Brandt chaired his own commission and established the Commission on Global Governance. Maurice Strong was Secretary-General of the Stockholm and Rio Conferences, first Secretary-General of the UN Environment Programme, and served on the Brundtland Commission and the Commission on Global Governance. Sridath Ramphal was Secretary-General of the Commonwealth, a member of the Brandt, Palme, Brundtland, and South Commissions, and co-chaired the Commission on Global Governance. The focus on eminent individuals from government and the predominance of the same individuals across commissions calls into question whether "freshness and innovation in global governance," one of the aims of the Commission on Global Governance, can be achieved with the model described here.

The Growth of Transboundary Alliances

Even as these global commissions were being established, attempts to resolve contentious issues were increasingly shifting from state-centred efforts to those involving non-governmental actors, with civil society organisations as well as corporate and other market actors helping to create private or semi-public regulation. "Multilateralism," understood as relations among states, was increasingly being re-constituted to become "multi-stakeholder."[56] Although earlier efforts were among representatives of governments, a shift was taking place in the plurality of actors engaged in the process of influencing both the global and the national policy process. In this section, we describe how non-governmental actors have gained growing representation in international meetings and institutions and have even become instrumental in establishing international regimes that provide regulation or norm setting where governments were either unwilling or absent.[57]

Civil society organisations became adept at influencing official agendas.

Over the past decade and a half, the range, diversity, and awareness of issues that transcend national borders and disciplinary boundaries have expanded: the polarisation of wealth and the increasing disparities within and between nations; social inequities; the adverse impacts of the present trading regimes; debt; and the politics of natural resource use and defence spending. Not surprisingly, this complexity has led to more transboundary networks, national and local movements, and international organisations. These networks have pushed the boundaries of the conventional agendas of international intervention from those minimal processes of consultation to transnational networks playing a role in agenda setting, influencing official discourse and specific policies, and changing the behaviour of governments.[58] In the process, transnational alliances have played a growing role in redefining who should sit at the table and what the agenda should be.

Of considerable relevance to the WCD is the history of civil society organising to increase multilateral development banks' (MDBs) accountability to civil society in both borrower and donor countries.[59] From the early 1980s, this "MDB

Campaign" has deployed a range of tools, including media education, protests in donor and borrower countries, lobbying key officials (senior politicians, administrators, and World Bank Executive Directors), and hearings in European parliaments and the U.S. Congress. Instrumental to the successes achieved by the campaign have been the links between NGOs in donor and borrower countries. Northern NGOs used their leverage and advocacy with donor governments to push for reform at the MDB's. Advocacy for this reform was based on local project information from Southern NGOs and the political legitimacy their participation provided.[60] The resultant reforms were intended to encourage borrower governments to respond to social and environmental concerns and create political space for Southern NGOs to engage their own governments.[61]

The independent review of the Sardar Sarovar Dam was central to civil society's campaign against multilateral development banks.

A central moment in this campaign was the World Bank's establishment of an independent review of the Sardar Sarovar Dam Project on the Narmada River in India.[62] Faced with extreme pressure from the anti-dam movement in India and their partners in Washington, considerable scrutiny from the media and the U.S. Congress, and fading international credibility, the World Bank appointed Bradford Morse, who had recently stepped down as head of the UN Development Programme, to assess the project with particular attention to resettlement and amelioration of environmental aspects.[63] The Morse Commission legitimised central elements of the NGO coalition's critique of the project, which lent the campaign greater momentum and credibility. The Morse Commission was a precursor of the World Commission on Dams, in terms of its origins in the struggle by citizens' groups and the fact that it was appointed by the World Bank to provide an independent view on a dams-related conflict.

Over the 1980s and 1990s, the MDB Campaign left a considerable imprint on the structure and functioning of the World Bank.[64] The institution adopted policies on involuntary resettlement, indigenous peoples, and environmental assessments. In addition, the World Bank established a new information disclosure policy and instituted an Inspection Panel as an appeals mechanism against World Bank projects in direct response to campaign efforts. Few would argue that practice on the ground has measured up to the promise or that the underlying mission and mandate of MDBs have been transformed. Yet few would also dispute that the campaign has led to better mechanisms of accountability over MDBs.

In addition to propelling reform at existing institutions, civil society organisations have become increasingly adept at participating in and shaping the formation of new international regimes. In the environmental arena, the 1987 negotiation of the Montreal Protocol on Substances that Deplete the Ozone Layer marked the beginning of NGO influence.[65] A dozen or more industry groups represented the private sector. Although obstructionist at first, a shift toward co-operation by some companies was critical to success. Only two or three NGO groups followed these highly technical and somewhat esoteric, if critical, negotiations. However, these NGOs played important roles in educating the public, building consumer pressure through threat of boycotts, and lobbying governments during negotiations. By the conclusion of the process, according to the chief U.S. negotiators, proposals were not only discussed among country negotiators, but also with industry and environmental groups.[66]

It was in the subsequent, and more visible, negotiations on a global climate treaty in the run-up to the Earth Summit that NGOs moved from "out of the hallways" to "around the table."[67] NGOs served on delegations, were sometimes allowed into meeting rooms, and played a major role in defining the negotiating strategy of the delegations from some small island nations under threat from climate change. It was NGOs who put the issue on the international agenda, began the process of forging a scientific consensus on the need to address the problem, proposed a structure for the treaty, and mobilised public pressure for action.[68] In the build-up to the Earth Summit, NGOs from North and South were organised under the umbrella of the Climate Action Network. The Network's immediate goal—a treaty to be signed at Rio in 1992—tended to overshadow and pre-empt discussion of the broader socio-economic context of the climate issue, to the frustration of some Southern NGOs.[69] There was much debate within the Network on the relative merits of a focussed, if narrow, strategy versus one that addressed the larger political issues that shaped the climate debate. Indeed, as Keck and Sikkink (1998) have suggested, this transnational advocacy network, which involved people in structurally unequal positions, became a site "for negotiation over which goals, strategies and ethical understandings are compatible."[70] Thus, the Network not only formed the basis for sharing information and co-ordinating strategy, but also provided a framework for negotiating disputes across North-South lines in a foreshadowing of similar debates within governments. In sum, the climate negotiations comprehensively demonstrated the deep engagement and influence of NGOs in setting agendas and shaping processes that establish global regimes.

As with the ozone negotiations, the private sector was also active in the climate negotiations and at the national level through an umbrella Global Climate Coalition. Industry groups actively sought to undermine the scientific consensus on climate change, influence public opinion on both the science and economic costs of mitigation policies, and engage with delegations of countries sympathetic to their views.[71]

Increasingly, the private sector has also organised itself into broader, less issue-specific networks based on dialogue, rather than advocacy. One such example, the Business Council for Sustainable Development (later to become the World Business Council on Sustainable Development—WBCSD), formed in preparation for the Earth Summit.[72] The WBCSD is a coalition of 50 international companies united by a shared commitment to sustainable development.[73] The organisation pursues this goal via the three pillars of economic growth, environmental protection, and social equity. Members are drawn from more than 30 countries and 20 major industrial sectors.[74] The WBCSD represents an evolution from the more partisan and narrowly-focussed issue networks that characterised the ozone and climate negotiations.

Another significant recent mobilisation, around the issues of trade and investment, comes from the same source as the NIEO and the Brandt and South Commissions—a concern with economic justice. However, although governments drove the economic justice commissions of the 1970s and 1980s, the more recent expressions of concern with North-South issues have been raised by social movements. Economic justice campaigns differ from advocacy around global environmental treaties, which have been dominated by technocratic NGOs. The breadth of organising around issues of trade and economic justice has been considerable. Negotiations around the North American Free Trade Agreement (NAFTA) stimulated civil society networks in the United States and Mexico on a large scale. The efforts at political organising, the corresponding political clout generated, and the associated research on labour and environmental effects of a free trade agreement contributed to the preparation of side-agreements to NAFTA on environment and labour.[75]

The private sector has increasingly organised into broad networks based on dialogue rather than advocacy.

Through rapid mobilisation and advances in telecommunications, civil society groups were able to halt in 1998 the "Multilateral Agreement on Investment" (MAI), which was being negotiated behind closed doors at the Organisation for Economic Co-operation and Development (OECD).[76] A broad and deep movement was mobilised around the proposed increase in the rights of corporations, and an absence of countervailing obligations. In addition, civil society objected vehemently to the lack of scope for input by public interest groups and the lack of information about the process available to the public. The MAI campaign not only demonstrated the capability of civil society to mobilise rapidly, but also that processes of global governance without public input and participation had little legitimacy and even less chance of success.

In short, although the phenomenal growth of communications and movement across national boundaries has undoubtedly stimulated transnational finance and trade resulting in the marginalisation of entire countries, cultures, and communities,[77] the communications revolution has also created new possibilities for regional and global networking—a parallel globalisation of mobilisation, aimed at expanding the participation of hitherto excluded voices in the centre of political decision-making. New transnational communities are evolving a globalisation from below. In many cases, we are witnessing the emergence of global advocacy networks and what some have even called "transnational social movements."[78]

The rise in transnational networks brings with it a host of questions about the future of international politics and the role of civil society.[79] Some see this vast outpouring of institutional and political innovation represented by transboundary alliances as messy, chaotic, and fragmented. Those sympathetic to the overall objectives argue for greater central co-ordination, more coherent, if interrelated, strategies and a clearer set of institutions guiding transboundary civil society processes. In response, an important body of analysis examines the internal dynamics of transboundary networks, with particular emphasis on the significant differences between Northern and Southern participants. Indeed, the representativeness and accountability of NGOs themselves is a concern. More sceptical voices offer the possibility that civil society mobilisation will lead to excessive international pluralism, creating a logjam of interests and rendering political systems unworkable.[80] While these are valid concerns, the examples discussed here suggest that networks of civil society organisations have, at a minimum, acted as "tugboats in international channels."[81] They are increasingly capable of influencing, or even setting, agendas. Civil society actors are not simply bringing critical concerns from the grassroots to the table, but are actively participating in the process of governing.

The Rise of Multi-Stakeholder Processes and the Challenge of Compliance

The emergent forms of conflict and dialogue described above among governments and international organisations, social movements, NGOs, and the private sector is beginning to redefine the form and content of global governance. At the international level, the consent of other governments is no longer enough for governments to secure legitimacy and act unhindered. In the era of economic globalisation, governments have become weaker in controlling capital flows and corporate investment, with a consequent loss of sovereignty over decisions and processes from outside their boundaries that impact the lives of the citizenry within.[82] Meanwhile, governments are proving inadequate in ensuring governance that facilitates and respects the deepening of democracy and justice.[83] In fact, governments can be causes of continuing social, cultural, and ecological costs, leading to disillusionment with their performance and role.[84] How can diverse actors creatively explore the complex process of governance without government? And what kinds of organisations or organisational arrangements are necessary to manage and administer diverse social, economic, and ecological systems? Can transboundary networks, then, creatively occupy this vacuum and participate in the definition of new norms of global governance?[85]

States have become weaker in controlling capital flows and corporate investment,with a consequent loss of sovereignty.

For their part, the private sector has long been perceived as an influential and even essential voice in policy formulation, perhaps because of state dependence on private sector profitability for jobs and taxes, and, therefore, for domestic and international stability.[86] As the examples above suggest, civil society voices are also increasingly indispensable to legitimate process and are well placed to fill the space occupied by retreating governments. And yet, there are few models of global governance that provide channels for direct participation of all these groups.[87]

In this transformed global context of multiple players, there has been increasing attention to and interest in structuring international governance processes around the participation of multiple actors—multi-stakeholder processes. Much of the prior experience with MSPs has been at the national level. For example, in Canada MSPs have been organised at the river basin level and on the intersections between the environment and the economy at the provincial and national levels.[88]

A recent effort at deriving a systematic framework for MSPs focussed at the global level suggests that they have the potential to promote better decision-making and to increase the chances of implementation.[89] In this view, MSPs convene actors with a breadth of perspectives, many of whom are often left out of state-centred decision-making process. Moreover, MSPs help integrate these diverse subjective viewpoints, resulting in better decisions. Finally, since inclusion in decision-making processes helps build constituencies for implementation, decisions are more likely to be carried out. Thus, MSPs, when well designed and implemented, provide for inclusion of previously excluded views, enrich decision-making, and increase the probability of implementation.

Multi-stakeholder processes help integrate diverse viewpoints.

It is important to recognise that the term "MSP" allows for considerable heterogeneity. Thus, these processes may vary in the objectives they promote, which include informing decision-making, providing an opportunity for dialogue, and monitoring implementation. MSPs can also range widely in their definition and categorisation of stakeholders. Although some processes use a simple trilateral formulation of civil society, private sector, and government, others have more fine-grained categories, such as the nine categories developed for Agenda 21 of the 1992 Earth Summit. Moreover, the scope and timeline of MSPs must be tailored to the issue at hand.

Although there is growing enthusiasm for such processes, the underlying assumptions about what and how they contribute to policymaking also differ widely. In a study of global public policy networks, a concept not dissimilar to MSPs, Reinicke and his colleagues suggest that these processes can bridge both a "participatory gap" that precludes effective participation in decision-making and an "operational gap" in information, knowledge, and tools in a world of economic and political liberalisation.[90]

Even if these gaps are bridged, however, a critical determinant of MSPs' effectiveness is how the outcomes of such dialogues link to decision-making. MSPs often lack formal authority or linkage to decision-making and, other than seeking to build stakeholder buy-in to a process, have few tools with which to implement their results. By contrast, governments still retain various mechanisms for policy enforcement and there are very few transnational social movements that significantly influence state behaviour or corporate or financial capital. Indeed, the lack of ability to ensure compliance is one reason why some civil society groups mistrust MSPs.

Despite the lack of formal mechanisms of compliance, non-governmental actors can play a useful role in forging good governance through the promotion of norms. Indeed, there is a long history of effective social change brought about as norms emerge, gain mainstream currency, and cascade through society, in part as nations and societies re-fashion political identities around these norms. Over time, norms may be internalised in political and institutional systems through laws and bureaucratic regulations.[91] This process is well illustrated by the progressive adoption of women's suffrage and, more recently, the anti-personnel land mine ban. In the case of the latter, a working partnership between the Canadian government and a coalition of over 350 humanitarian and arms-control NGOs from 23 countries acted as "norm entrepreneurs."[92] Over time, social pressures of identity have brought about an emulation or a cascade effect.[93] Norm creation and promotion offers a complementary approach to traditional forms of compliance through legal sanction.

The WCD in Historical Context

The concerns, emergence, structure, and functioning of the WCD draws from diverse strands in the history of global governance—global commissions, growing civil society influence, and the growing acceptance of multi-stakeholder processes. In terms of the concerns that motivated it, the WCD is located at the intersection of the debates over environment and development, on the one hand, and economic justice on the other, that motivated the series of UN-sponsored commissions of the 1970s and 1980s. Built as engineering marvels to provide water and electricity and to control floods, dams have come under increasing criticism for their destruction of the environment and communities and for their contribution to unfair development. These competing images of boon and blight place the debate at the locus of concerns of past commissions. The WCD was designed to illuminate the intersection of environment, development, and justice by shining a spotlight on the very process of planning, design, and implementation of large projects, to ask whether and how economic, social, cultural, and economic costs were the likely result of the existing framework.

That the WCD came into existence is due to a long history of struggle in many countries and by many community groups and NGOs.[94] The mobilisation around large dams suggests that local actors are thinking and acting at both local and global levels. While working for change in their own contexts, these actors have built political pressure for a process of change at the global level. The WCD is, then, firmly rooted in the growing strength of transnational networking and, in particular, in the campaign to reform the MDBs.

With regard to structure and functioning, the WCD marked a departure from past commissions based on eminent persons and exclusive processes. As Chapters 3 and 4 will make clear, the WCD was based on representation of stakeholders rather than on eminence, usually defined as a distinguished career in government. Moreover, this shift in emphasis toward representation was due to the forceful role that civil society organisations played in the formation of the WCD. In the move toward bringing diverse actors to the table, the WCD was emblematic of the growing interest in MSPs, with the potential for dialogue, development of constituencies for implementation, and norm creation that these new vehicles promise. In its operation, the WCD was firmly in the tradition of past commissions and processes in seeking to forge an independent path. Moreover, in its attempts at establishing an inclusive process, it took its cue from the Brundtland Commission's efforts to solicit a broad range of views, as Chapters 5 and 6 describe. However, it went further than past processes in its explicit commitment to transparency as a means of supporting its claims to legitimacy.

In bringing all these elements together and in applying them to a focussed issue area that had been the subject of contention, the WCD had no direct comparators. Although rooted in historical context, it was an experiment that sought to take significant steps beyond the collective past of global governance. The remaining chapters tell the story of how far the experiment succeeded in meeting the expectations it generated.

Endnotes

1. The literature on the current phase of economic globalisation is vast. The following books provide a useful introduction to its depth and complexity and are relevant for the discussions here: Ulrich Beck, *What is Globalisation?* (Cambridge: Polity Press, 2000); J. Mittelman, *The Globalization Syndrome. Transformation and Resistance* (Princeton, NJ: Princeton University Press, 2000); P. Hirst and G. Thompson, *Globalisation in Question: The International Economy and the Possibilities of Governance* (Cambridge: Polity Press, 1999); *Globalisation and the Welfare State* (Cheltenham, UK: Edward Elgar, 1999); and Thomas L. Friedman, *The Lexus and the Olive Tree: Understanding Globalization* (New York: Farrar, Straus and Giroux, 1999).
2. This architecture was pre-dated by the formation of the League of Nations after World War I. The League of Nations set the stage for global governance on which the post World War II institutions were built.
3. The United Nations and associated organisations were conceptualised and given a firm identity during conversations at Bretton Woods in July 1944 and Dumbarton Oaks in August 1944.
4. Andrew E. Rice and Cyril Ritchie. 1995. "Relationships between International Non-governmental Organizations and the United Nations." Originally published in *Transnational Associations* 47(5): 254-265. Online at: www.uia.org/uiadocs/unngos.htm (28 September, 2001).
5. United Nations. 21 December 1984. Document ST/SGB/209, Secretary-General's Bulletin. Online at: www.un.org/partners/civil_society/document/sgb_209.htm (28 September 2001).
6. Global Policy Forum. "NGO Review – November 1996. An Analysis by NGLS – the UN's Non-Governmental Liaison Service." Online at: www.globalpolicy.org/ngos/analysis/nglsnv96.htm (28 September 2001).
7. E-mail communication with IUCN staff, October 2001.
8. Global Policy Forum. 1998. "The Coffers are Not Empty: Financing for Sustainable Development and the Role of the United Nations." Online at: www.igc.org/globalpolicy/socecon/global/paul.htm (28 September 2001).
9. Samir Amin, ed. *Maldevelopment: Anatomy of a Global Failure* (United Nations University Press, 1990). Online at: www.unu.edu/unupress/unupbooks/uu32me/uu32me06.htm (28 September 2001).
10. Rachel Carson, *Silent Spring* (Boston: Houghton Mifflin, 1962).
11. Ramachandra Guha, *Environmentalism: A Global History* (New York: Longman, 2000).
12. Dick Richardson and Chris Rootes, *The Green Challenge: The Development of Green Parties in Europe* (London & New York: Routledge, 1995).
13. The Club of Rome is a global think tank and research centre. It brings together scientists, economists, businesspersons, international civil servants, and present and former heads of state from across the world who are convinced that a humane future can be shaped by bold collective research and action. Club of Rome website, www.clubofrome.org/ (28 September 2001).
14. Donella H. Meadows et al., *The Limits to Growth* (New York: Universe Books, 1972).
15. The Ecologist. 1972. "Blueprint for Survival," *The Ecologist* 2(1): 1-43.
16. Ramchandra Guha and Juan Martinez-Alier, *Varieties of Environmentalism. Essays North and South* (New Delhi: Oxford University Press, 1997).
17. Guha, 2000.
18. "Official Aid—A Brief History," *New Internationalist* (November 1996). Online at: www.oneworld.org/ni/issue285/history.html (28 September 2001).
19. Ken Conca. September 1995. "Greening the United Nations: Environmental Organizations and the UN System," *Third World Quarterly* 16(3): 441-57.
20. Tanja Brühl and Udo E. Simonis. 2001. "World Ecology and Global Environmental Governance," Berlin: WZB paper (FS II 01-402). Online at: http://skylla.wz-berlin.de/pdf/2001/ii01-402.pdf (28 September 2001).
21. Rice and Ritchie, 1995.
22. Conca, 1995.
23. United Nations Environment Programme, *In Defence of the Earth: The Basic Texts on Environment: Founex • Stockholm • Coycoyoc*, Executive Series 1 (Nairobi: United Nations Environment Programme, 1981).
24. David Rockefeller, credited with the idea of the Trilateral Commission, quoting from his own remarks in March 1972 during a speech on the occasion of the Trilateral Commission U.S. Group's 25th anniversary, 1 December 1998. Online at: www.trilateral.org/nagp/regmtgs/98/1201rockflr.htm (28 September 2001).
25. Trilateral Commission website, www.trilateral.org/about.htm (28 September 2001).
26. In recent times, the Commission's composition has broadened geographically to embrace the Asia-Pacific region, Mexico, and various European countries with the expansion of the EU. Its composition has also broadened to include media, academia, labour unions, and NGOs. Trilateral Commission website, www.trilateral.org/about.htm (28 September 2001).
27. Independent Commission on International Development Issues, *North-South: A Program for Survival. The Report of the Independent Commission on International Development Issues under the Chairmanship of Willy Brandt* (Cambridge, MA: MIT Press, 1980).
28. Independent Commission on International Development, *North-South: A Program for Survival. The Report of the Independent Commission on International Development Issues under the Chairmanship of Willy Brandt* (Cambridge, MA: MIT Press, 1980).
29. Leonard Downie Jr., "New Ideas on Global Co-operation Unveiled," *The Washington Post* (19 December 1979); and "The Brandt Commission's Proposals for Survival," *The Economist* (22 December 1977).
30. Downie Jr., 1979.
31. Sustainable use is defined as "...the integration of conservation and development to ensure that modifications to the planet do indeed secure the survival and well-being of all people." IUCN/UNEP/WWF/FAO/UNESCO, *World Conservation Strategy: Living Resource*

Conservation for Sustainable Development (Gland, Switzerland: IUCN, 1980).

32. In between the Brandt and Brundtland Commissions, the Palme Commission, or the Independent Commission on Disarmament and Security Issues, was established under UN auspices to chart a course toward disarmament. The Palme Commission completed its work between 1980 and 1982, and, as with past commissions before it, was composed of statesmen and women chosen for their "broad political experience." While the Commission did rely heavily on submissions by experts, it did also attempt to maintain contact with NGOs, and dedicated one secretariat staff member to this task (Rockefeller Brothers Fund website, www.rbf.org/pws/palme1.html, 25 June 2001). Since the Palme Commission focussed on security issues, and not on either economic or environmental themes, the focus of this review, we do not address it further here.
33. World Commission on Environment and Development, *Our Common Future: The World Commission on Environment and Development* (Oxford: Oxford University Press, 1987).
34. World Commission on Environment and Development, 1987.
35. Pascale D. Morand Francis, *Geneva at the International Crossroads of Environment and Development* (Geneva: Swiss Federal Department of Foreign Affairs, 1998). Online at: http://geneva-international.org/GVA3/WelcomeKit/Environnement/chap_5.E.html (28 September 2001).
36. D. Reid, *Sustainable Development: An Introductory Guide* (London: Earthscan, 1995). For a history, see Richard Sandbrook, "From Stockholm to Rio - Earth Summit 1992," *The United Nations Conference on Environment and Development*. J. Quarrie, ed. (London: Regency Press, 1992). For a historical critique of the term "sustainable development" and the Rio Summit, see Wolfgang Sachs, *Planet Dialectics* (London: Zed Books, 1999).
37. J. Gardner and M. Roseland. 1989. "Thinking Globally: The Role of Social Equity in Sustainable Development," *Alternatives* 16(3): 26-34.
38. South Centre website, www.southcentre.org (28 September 2001).
39. South Commission, *The Challenge to the South: The Report of the South Commission* (Oxford: Oxford University Press, 1990).
40. South Commission, 1990.
41. Brühl and Simonis, 2001.
42. Peter Haas, Marc Levy and Edward Parson. October 1992. "Appraising the Earth Summit: How Should we Judge UNCED's Success?" *Environment*, 34(8): 6-11, 26-33.
43. Mark Valentine, "Twelve Days of UNCED," U.S. Citizens Network, Tides Foundation, July 2, 1992.
44. Lamont C. Hempel, *Environmental Governance: The Global Challenge* (Washington, DC: Island Press, 1996).
45. However, inequitable access to the electronic world remained an outstanding issue. Many groups and movements who were not linked into these networks sought to moderate claims that communication technology is "emboldening people in ways that have profound implications for the process of democratisation." See Sheldon Annis, "Giving Voice to the Poor," *Foreign Policy*, vol. 84 (Fall 1991): 93-106.
46. For a critical and exhaustive review of the Earth Summit, see Pratap Chatterjee and Matthais Finger, *The Earth Brokers: Power, Politics and World Development* (New York: Routledge, 1994).
47. Matthias Finger, "Environmental NGOs in the UNCED Process," in *Environmental NGOs in World Politics*. Finger and Princen, eds. (New York: Routledge, 1994).
48. UNCSD website, http://www.un.org/esa/sustdev/csdgen.htm (28 September 2001).
49. Commission on Global Governance, *Our Global Neighbourhood: The Report of the Commission on Global Governance* (Oxford: Oxford University Press, 1995). Online at: www.cgg.ch/ (28 September 2001).
50. Commission on Global Governance, 1995.
51. Although established in 1993, the World Commission on Forests and Sustainable Development was formally launched in 1995.
52. WCFSD website, http://iisd1.iisd.ca/wcfsd (28 September 2001).
53. Independent World Commission on the Oceans, *The Ocean, Our Future* (Cambridge: Cambridge University Press, 1998).
54. World Water Council, *World Water Vision: Making Water Everybody's Business* (London: Earthscan, 2000). Online at: www.worldwatervision.org/reports.htm (28 September 2001).
55. "Old Water in a New Bottle: World Water Vision is Chronically Short-sighted." Written by International Rivers Network (USA), International Committee on Dams, Rivers, and People, and Both Ends (Netherlands), and endorsed by 16 non-governmental groups from Brazil, England, India, Nepal, Pakistan, Slovakia, South Africa, Switzerland, and Thailand, 17 March 2000.
56. Ruggie describes conventional multilateralism as "an institutional form that co-ordinates relations among three or more states on the basis of generalized principles of conduct." J.G. Ruggie. 1992. "Multilateralism: The Anatomy of an Institution." *International Organization* 46(3): 561-598.
57. While non-state actors have gained growing relevance on the international scene, there is also a long history of significant past international interventions by NGOs. Charnovitz traces international activity by NGOs back to 1775. Since then, NGOs have played a significant role in the abolition of slavery, free trade, human rights, women's rights, and a host of other issues. Steve Charnovitz. Winter 1997. "Two Centuries of Participation: NGOs and International Governance," *Michigan Journal of International Law* 18(2): 183-286. Also see UNRISD, *Civil Society, NGOs, and Social Development: Changing the Rules of the Game*, Occasional Paper No. 1, January 2000.
58. For a useful analytical framework along these lines to assess the impact of transnational networks, see M. Keck and K. Sikkink, *Activists across Borders: Advocacy Networks in International Politics* (Ithaca: Cornell University Press, 1998).

59. This struggle has spawned a voluminous commentary. For a small but insightful selection, see Barbara Bramble and Gareth Porter, "Non-Governmental Organizations and the Making of US International Environmental Policy," in *The International Politics of the Environment.* A. Hurrell and B. Kinsbury, eds. (Oxford: Clarendon Press, 1992); Bruce Rich, *Mortgaging the Earth: The World Bank, Environmental Impoverishment, and the Crisis of Development* (Boston: Beacon Press, 1994); Jonathan Fox and L. David Brown, eds. *The Struggle for Accountability: The World Bank, NGOs and Grassroots Movements* (Boston: MIT Press, 1998); Robert Wade, "Greening the Bank: The Struggle over the Environment, 1970-1995," in *The World Bank: Its First Half-Century.* Devesh Kapur, John P. Lewis, and Richard Webb, eds. (Washington: Brookings Institution, 1997); Paul J. Nelson, *The World Bank and Non-Governmental Organizations: The Limits of Apolitical Development* (New York: St. Martin's Press, 1995).
60. As Fox and Brown, 1998, note, using the leverage of the Bank for reformist ends did arouse sovereignty concerns among Southern groups. Moreover, whether Southern groups were equal partners with a voice in strategic decisions, or whether their role was limited to project information and political cover was an open question during the early years of the campaign. See Navroz K. Dubash, "The Birth of an Environmental Movement: The Narmada Valley as Seed-Bed for Civil Society in India." A.B. Thesis, Woodrow Wilson School of Public and International Affairs, Princeton University, April 1990.
61. Fox and Brown, 1998, pp. 6-7.
62. For the World Bank's role in the Narmada dam projects, see William Fisher, ed. *Toward Sustainable Development? Struggling over India's Narmada River* (ME Sharpe, 1995).
63. Bradford Morse and Thomas Berger, *Sardar Sarovar: The Report of the Independent Review* (Ottawa: Resource Futures International, 1992). Also see Smitu Kothari, "The Narmada Movement, Transnational Alliances, and Democracy," in *Transnational Civil Society*. Kathryn Sikkink, Sanjeev Khagram, and James Riker, eds. (Minneapolis: Minnesota University Press, forthcoming 2001).
64. Fox and Brown, 1998.
65. Discussion of the Montreal Protocol is drawn from Richard Benedick, *Ozone Diplomacy: New Directions in Safeguarding the Planet* (Cambridge, MA: Harvard University Press, 1991).
66. Benedick, 1991, p. 204.
67. Jessica T. Mathews, "Power Shift," *Foreign Affairs* (January/February 1997): 50-51.
68. In addition, during negotiations, NGOs lubricated dialogue by providing back channels for communication and produced a daily newspaper, ECO, which was used by negotiators as a credible information source and a way of testing ideas to break deadlocks. Mathews, 1997.
69. Navroz Dubash and Michael Oppenheimer, "Modifying the Mandates of Existing Institutions: NGOs," in *Confronting Climate Change: Risks, Implications and Responses.* Irving Mintzer, ed. (Cambridge: Cambridge University Press, 1992).
70. Keck and Sikkink, 1998.
71. David L. Levy and Daniel Egan. 1998. "Capital Contests: National and Transnational Channels of Corporate Influence on the Climate Change Negotiations," *Politics and Society* 26(3): 335-359.
72. As Errol Meidinger has usefully observed about environmental regulation, "Private organizations have recently established numerous programs aimed at improving the environmental performance of industry. Many of the new programs seek to define and enforce standards for environmental management, and to make it difficult for producers not to participate in them. They claim, explicitly and implicitly, to promote the public interest. They take on functions generally performed by government regulatory programs, and may change or even displace such programs. Private environmental regulatory programs thus have the potential to significantly reshape domestic and international policy institutions by changing the locus, dynamics, and substance of policy making." E.E. Meidinger. 2000. "'Private' Environmental Regulation, Human Rights and Community," *Buffalo Environmental Law Journal* 7(1).
73. The WBCSD was formed in January 1995 through a merger between the Business Council for Sustainable Development (BCSD) in Geneva and the World Industry Council for the Environment (WICE), an initiative of the International Chamber of Commerce (ICC), in Paris. Those two parent bodies had been active in evolving business's response to the challenges arising from the Earth Summit in Rio de Janeiro in 1992.
74. For full list of members and details of policy oriented activities, as well as a full profile, see World Business Council for Sustainable Development website, www.wbcsd.ch (28 September 2001).
75. Raul Hiojosa-Ojeda, "Institution Building within the NAFTA Context: An Evaluation of Policy Initiations from the Transnational Grassroots," Berkeley Roundtable on the International Economy, Working Paper 95, July 1999. Also see Mathews, 1997.
76. See UNCTAD, "Lessons from the MAI", *UNCTAD Series on Issues in International Investment Agreements* (New York and Geneva: UNCTAD, 1999); D. Henderson, *The MAI Affair: A Story and its Lessons* (London: Royal Institute for International Affairs, 1999); *Lokayan Bulletin*, Special Issue on the MAI, Vol. 14, No. 1, 1998.
77. Manuel Castells has called those who are marginalised by dominant processes of economic development (particularly in this age of globalisation) as "structurally irrelevant." Manuel Castells, *The Information Age: Economy, Society and Culture. Vol. 1: The Rise of the Network Society* (Cambridge, Massachusetts: Blackwell, 1996).
78. Donatella Della Porta, Hanspeter Kriesi and Dieter Rucht, eds. *Transnational Social Movements* (London: Macmillan, 1999). Also see Jackie Smith, Charles Chatfield, and Ron Pagnucco, eds. *Transnational Social Movements in Global Politics* (Syracuse: Syracuse University Press, 1997).
79. The discussion in this chapter is drawn from a range of recent articles on the role of civil society in global environmental politics: Keck and Sikkink, 1997; Lisa Jordan and Peter van Tujl. 2000. "Political Responsibility in Transnational NGO Advocacy," *World Development* 28(12): 2051-2065; Michael Edwards, *Future Positive: International Co-operation in the 21st Century* (London: Earthscan, 2000); Michael Edwards, "NGO Rights and

Responsibilities." *New York Times* op ed piece, 2001. Online at: www.futurepositive.org/NYT.doc (28 September 2001); Robert O'Brien et al., *Contesting Global Governance: Multilateral Economic Institutions and Global Social Movements* (Cambridge: Cambridge University Press, 2000).

80. Mathews, 1997. Yet others go further to argue that many NGOs are politically naïve, as well as unaccountable and non-representative, and are manipulated by dominant powers into accepting marginal changes as progress. See, for example, Hugo Slim, "To the Rescue: Radicals or Poodles?" *The World Today*, vol. 53 (Aug/Sept 1997): 209-212.

81. Ann Marie Clark. Winter 1995. "Non-governmental Organisations and their Influence on International Society." *Journal of International Affairs* 48(2): 507-525.

82. P. Hirst and G. Thompson, *Globalisation in Question: The International Economy and the Possibilities of Governance* (Cambridge: Polity Press, 1999).

83. S. Strange, *The Retreat of the State* (Cambridge: Cambridge University Press, 1996).

84. For discussions of the role of governments, see Ian Shapiro and Casiano Hacker-Cordon, *Democracy's Edges* (Cambridge: Cambridge University Press, 1999); Robert A. Dahl, *On Democracy* (New Haven: Yale University Press, 1998); and Rajni Kothari, *Poverty, Human Consciousness and the Amnesia of Development* (London: Zed Books, 1993).

85. For a more detailed analysis of the processes of "governance without government," see James Rosenau and Ernst-Otto Czempiel, eds. *Governance Without Government: Order and Change in World Politics* (New York: Cambridge University Press, 1992).

86. Levy and Egan, 1998.

87. An exception is the International Labour Organization, which gives voting rights to labour unions and the private sector, in addition to governments. See P.J. Simmons, "Learning to Live with NGOs", *Foreign Policy* (Fall 1998): 82-96.

88. The Canadian experience is particularly relevant to the WCD because the primary facilitator for the WCD process, Prof. Anthony Dorcey, drew much of his experience from practice in Canada. See, for example, A.H.J. Dorcey, "Collaborating Towards Sustainability Together: The Fraser Basin Management Board and Program," in *Practising Sustainable Water Management: Canadian and International Experiences*. D. Shrubsole and B. Mitchell, eds. (Cambridge, Ontario: Canadian Water Resources Association, 1997); and A.H.J. Dorcey, L. Doney and H. Rueggeberg, "Public Involvement in Government Decision-Making: Choosing the Right Model." B.C. Round Table on the Environment and the Economy (Victoria, 1994).

89. Minu Hemmati et al., *Multi-Stakeholder Processes for Governance and Sustainability Beyond Deadlock and Conflict* (London: Earthscan, 2001). Online at: www.earthsummit2002.org/msp/ (28 September 2001).

90. Wolfgang Reinicke, Francis Deng et al., *Critical Choices: The United Nations, Networks and the Future of Global Governance* (Ottawa: International Development Research Centre, 2000).

91. Martha Finnemore and Kathryn Sikkink. Autumn 1998. "International Norm Dynamics and Political Change." *International Organization* 52 (4): 887-917.

92. Simmons, 1998.

93. Richard Price. Summer 1998 "Reversing the Gun Sights: Transnational Civil Society Targets Land Mines." *International Organization* 52(3): 613-644.

94. See Sanjeev Khagram, "Toward Democratic Governance for Sustainable Development: Transnational Civil Society Organizing around Big Dams," in *The Third Force: The Rise of Transnational Civil Society*. Ann Florini, ed. (Washington, DC: Carnegie Endowment for International Peace, 2000).

Chapter 3

The Origins of the World Commission on Dams

In this chapter, we detail the immediate origins of the World Commission on Dams. We locate it both in the growth of local struggles against the adverse social, economic, and ecological impacts of dams and in the growing pressure to define global norms for harnessing and managing water. These developments compelled the World Conservation Union (IUCN) and the World Bank to organise a meeting of representatives from different sides of the dams debate, where participants decided to set up the Commission. We trace the many difficult steps necessary to create a body and a process that would satisfy all the stakeholders. This narrative illustrates the process and the challenges of constituting a multi-stakeholder institutional response to a highly contentious national and international issue.

Seeds of Dissent

It is difficult to pinpoint a single defining moment that led to the birth of the WCD. Over the past three decades, with increasing frequency and intensity, questions have been raised about the economic viability and the social, cultural, and environmental costs of large dams. It is important to note that, most often, in the absence of local mobilisation and social movements, information about planned dams is hard to come by and many poorly planned and implemented projects escape scrutiny.[1] But based on growing evidence of dams' negative impacts, protests and mobilisations have multiplied the world over. These protests have matured into sustained social movements that have effectively slowed down or stalled further work on proposed or ongoing dams. Among the more notable examples are the Bakun Dam in Malaysia, the Maan, Tehri, and Maheshwar Dams in India, and the Lesotho Highland Stage II Dam in Lesotho. In the case of proposed dams, such as the Arun III in Nepal, national mobilisation and intensive global campaigns have led to the cancellation of these projects. On the Koel-Karo and the Suvarnarekha Rivers in India, projects have been shelved after ground had been broken and significant infrastructure work had been completed. Even in the industrialised world—whether in the United States, Europe, or Japan—public opposition and the growing evidence of the adverse economic and ecological impacts have led to a rethinking of large dams as an option for irrigation and energy.[2] Additionally, social movements and their supporters have criticised the role of multilateral funding agencies such as the World Bank in the legitimation and construction of large dams.

A prominent example of this history of protest is the movement against dams on India's Narmada River. Domestic dissent to this project caused the World Bank to appoint an independent commission to assess these dams and subsequently an independent Inspection Panel to assess contentious projects.[3] The struggle against the Sardar Sarovar Project (SSP) in the Narmada Valley achieved several other global landmarks. It nudged the World Bank to review its central commitment to large dams, and its policies related to indigenous peoples and resettlement. It marked the first time that the Japanese government withdrew its direct and indirect support to a development project for environmental and human rights reasons.[4] It was the catalyst for formation of a remarkable national and transnational network of dam-affected people and their supporters.

In June 1994, on the 50th anniversary of the formation of the Bretton Woods institutions, a coalition of 326 social movements and non-governmental organisations (NGOs) from 44

Box 3.1

The Manibeli Declaration (excerpted)
Calling for a Moratorium on World Bank Funding of Large Dams
September 1994

WHEREAS:

1. The World Bank is the greatest single source of funds for large dam construction, having provided more than US$50 billion (1992 dollars) for construction of more than 500 large dams in 92 countries. Despite this enormous investment, no independent analysis or evidence exists to demonstrate that the financial, social and environmental costs were justified by the benefits realised.
2. Since 1948, the World Bank has financed large dam projects that have forcibly displaced on the order of 10 million people from their homes and lands. The Bank has consistently failed to implement and enforce its own policy on forced resettlement, first established in 1980.
5. The environmental and social costs of World Bank-funded large dams, in terms of people forced from their homes, destruction of forests and fisheries, and spread of waterborne diseases, have fallen disproportionately on women, indigenous communities, tribal peoples and the poorest and most marginalised sectors of the population. This is in direct contradiction to the World Bank's often-stated "overarching objective of alleviating poverty."
9. The Bank has even convinced governments to accept loans for large dams when more cost-effective and less destructive alternative plans existed.
10. The economic analyses on which the World Bank bases its decisions to fund large dams fail to apply the lessons learned from the poor record of past Bank-funded dams, underestimating the potential for delays and cost over-runs. Project appraisals typically are based on unrealistically optimistic assumptions about project performance, and fail to account for the direct and indirect costs of negative environmental and social impacts.

THEREFORE, the undersigned organisations:
CONCLUDE that the World Bank has to date been unwilling and incapable of reforming its lending for large dams; and CALL for an immediate moratorium on all World Bank funding of large dams including all projects currently in the funding pipeline, until:

1. The World Bank establishes a fund to provide reparations to the people forcibly evicted from their homes and lands by Bank-funded large dams without adequate compensation and rehabilitation. The fund should be administered by a transparent and accountable institution completely independent of the Bank and should provide funds to communities affected by Bank-funded large dams to prepare reparations claims.
2. The World Bank strengthens its policies and operational practices to guarantee that no large dam projects that require forced resettlement will be funded in countries that do not have policies and legal frameworks in place to assure restoration of the living standards of displaced peoples. Furthermore, communities to be displaced must be involved throughout the identification, design, implementation and monitoring of the projects, and give their informed consent before the project can be implemented.
3. The World Bank commissions, reviews, and implements the recommendations of an independent comprehensive review of all Bank-funded large dam projects to establish the actual costs, including direct and indirect economic, environmental and social costs, and the actually realised benefits of each project. The review must be conducted by individuals completely independent of the Bank without any stake in the outcome of the review.

Source: "Manibeli Declaration, Calling for a Moratorium on World Bank Funding of Large Dams," September 1994. Online at: www.irn.org/programs/finance/manibeli.shtml (28 September 2001).

countries around the world endorsed a statement calling for a moratorium on the World Bank's funding of large dams. This statement was named the Manibeli Declaration *(see Box 3.1)* in recognition of one of the first tribal villages that the Sardar Sarovar Dam on the Narmada River would submerge and one of the sites of sustained resistance to the dam. Crucially, one of the conditions for lifting the proposed moratorium was that the World Bank would set up an "independent comprehensive review of Bank-funded large dams projects to establish the actual costs, including the direct and indirect, economic, environmental, and social costs, and the actual realised benefits of each project." The Declaration went on to state that it was crucial that, "The review should evaluate the degree to which project appraisals erred in estimating costs and benefits, identify specific violations of Bank policies and staff members responsible, and address opportunity costs of not supporting project alternatives. The review must be conducted by individuals completely independent of the Bank without any stake in the outcome of the review."

The Operations Evaluation Department Review

Six months after the Manibeli Declaration, partly in response to criticisms of large dams and partly to deflect growing anger at the continuing involvement of the World Bank in these projects, the Operations Evaluation Department (OED) of the World Bank announced that it was "undertaking a review of World Bank-funded large dams in order to determine their development effectiveness."[5] As a senior official in OED noted, the World Bank's involvement in large dams had been "attracting a lot of controversy." Senior management felt pressed to address the question: "What is it about dams that causes so much concern and what should the Bank do about it?"[6]

The World Bank had originally envisaged the OED undertaking a two-phase study. Phase I was to be a desk review of experience with selected dams and Phase II was to be a more comprehensive study, involving field evaluations. The entire review was to be the World Bank's answer to its critics.

The first phase of the OED Review analysed the performance of 50 World Bank-funded dams. Its final report was sent to the World Bank President in mid-1996. The study stated that dams have contributed to economic development, including electricity generation capacity, flood control, and irrigation.[7] Based on an assessment against the standards in place at the time the project was implemented, the study found that resettlement was inadequate in half the dams funded, but that performance has improved over time. Performance on environmental grounds was deemed to be "mixed." The study further noted that while under prior social and environmental policies only 10 percent of the sample was unacceptable, had all projects been assessed under the new policies, 26 percent would have been unacceptable and 48 percent would have been judged "potentially acceptable." The report concluded that because the large majority of dams are yielding benefits that outweigh their costs, the World Bank should continue funding large dam projects with heightened attention to environmental and social policies. According to OED sources, the report was an internal, "relatively minor…desk study," considered inappropriate for public release.[8] NGOs criticised the report by arguing that the précis exaggerated the benefits of dams and that the full study was quoted selectively to justify the Bank's continued funding for more dams.

Although agreeing with OED's "positive conclusions," the Committee on Development Effectiveness of the Bank's Board of Executive Directors nevertheless, "…urged OED to ensure that Phase II reflects the views of civil society, including those of private investors and non-governmental organisations."[9] Although this recommendation did not call for wider participation in the second phase of the study, it is evident that the World Bank was aware of the potential lack of legitimacy and limited use of a study that reflected only its views on the contentious debate around large dams.

Aftermath of the OED Review

At the same time, the World Bank leadership was seeking a specific issue area for collaboration with the World Conservation Union (IUCN), with which it had cemented a partnership in 1994. "We had a partnership with IUCN but no substance," noted a World Bank official. "This [dams evaluation] showed up as something promising to focus on."[10] In this framework, the World Bank invited IUCN to co-organise a multi-stakeholder workshop to discuss the findings of the OED study and the goals and process for a more comprehensive second phase.[11] The meeting was proposed for Gland, Switzerland, in April 1997.

The run-up to the workshop illustrated only too well the urgent need for more constructive dialogue between the World Bank and its critics. Copies of the OED Review were leaked outside the World Bank and by September 1996 had galvanised anti-dam organisations and their supporters into action. On the eve of the Gland meeting, the International Rivers Network (IRN)—the California-based group at the forefront of research and co-ordination of individuals and organisations from around the world opposed to dams—published a lengthy and detailed critique of the OED Review. The response criticised the OED not only for underplaying the significance of the World Bank's own findings but also for its methodology and process.[12] IRN charged

Box 3.2

The Curitiba Declaration (excerpted)
Curitiba, Brazil, 14 March 1997

We, the people from 20 countries gathered in Curitiba, Brazil, representing organisations of dam-affected people and of opponents of destructive dams, have shared our experiences of the losses we have suffered and the threats we face because of dams. Although our experiences reflect our diverse cultural, social, political and environmental realities, our struggles are one. Our struggles are one because everywhere the people who suffer most from dams are excluded from decision-making. Decisions are instead taken by technocrats, politicians and business elites who increase their own power and wealth through building dams. Our shared experiences have led us to agree the following:

1. We recognise and endorse the principles of the 1992 'NGO and Social Movements Declaration of Rio de Janeiro' and the 1994 'Manibeli Declaration' on World Bank funding of large dams.
2. We will oppose the construction of any dam which has not been approved by the affected people after an informed and participative decision-making process.
3. We demand that governments, international agencies and investors implement an immediate moratorium on the building of large dams until:
   a. There is a halt to all forms of violence and intimidation against people affected by dams and organisations opposing dams.
   b. Reparations, including the provision of adequate land, housing and social infrastructure, be negotiated with the millions of people whose livelihoods have already suffered because of dams.
   c. Actions are taken to restore environments damaged by dams - even when this requires the removal of the dams.
   d. Territorial rights of indigenous, tribal, semi-tribal and traditional populations affected by dams are fully respected through providing them with territories which allow them to regain their previous cultural and economic conditions - this again may require the removal of the dams.
   e. An international independent commission is established to conduct a comprehensive review of all large dams financed or otherwise supported by international aid and credit agencies, and its policy conclusions implemented. The establishment and procedures of the review must be subject to the approval and monitoring of representatives of the international movement of people affected by dams.
   f. Each national and regional agency which has financed or otherwise supported the building of large dams have commissioned independent comprehensive reviews of each large dam project they have funded and implemented the policy conclusions of the reviews. The reviews must be carried out with the participation of representatives of the affected people's organisations.
   g. Policies on energy and freshwater are implemented which encourage the use of sustainable and appropriate technologies and management practices, using the contributions of both modern science and traditional knowledge. These policies need also to discourage waste and over consumption and guarantee equitable access to these basic needs.
4. The process of privatisation which is being imposed on countries in many parts of the world by multilateral institutions is increasing social, economic and political exclusion and injustice. We do not accept the claims that this process is a solution to corruption, inefficiency and other problems in the power and water sectors where these are under the control of the state. Our priority is democratic and effective public control and regulation of entities which provide electricity and water in a way which guarantees the needs and desires of people.

Source: "Declaration of Curitiba: Affirming the Right to Life and Livelihood of People Affected by Dams," 14 March 1997. Online at: www.irn.org/programs/curitiba.html (28 September 2001)

that a careful reading of the Review, which used three criteria—economic, social, and environmental—to evaluate the dams, clearly showed the extent to which the World Bank was trying to cover-up or minimise the implications of its own findings. Given that remedial action is rarely taken once a project is implemented, the Review showed that only 13 out of 50 dams funded by the World Bank were "acceptable." The IRN critique asserted that the World Bank believed, without justification, that remedial action was possible, that most dams were "potentially acceptable," and that they would be more acceptable than before, given increased safeguards on dam-related lending.

Based on the IRN critique, 49 NGOs from 21 countries wrote a collective letter to World Bank President James D. Wolfensohn demanding that the World Bank reject the conclusions of the OED Review: "Given the huge expense of large dams, the controversy over whether or not they are an effective means of achieving the benefits which their proponents claim for them, and the huge scale of their social and environmental impacts, a comprehensive, unbiased, and authoritative review of past World Bank lending for large dams is essential." Moreover, they argued that the review be undertaken by "a commission of eminent persons independent of the World Bank," which "must be able to command respect and confidence from all parties involved in the large dams debate."[13]

The NGO letter to the World Bank President built on prior civil society mobilisation against large dams. In March 1997, the First International Meeting of the People Affected by Large Dams, was held in Curitiba, Brazil. It was attended by activists and dam-affected people from more than 20 countries who shared the long history of the social, cultural, economic, and environmental problems associated with large dams, along with the undemocratic processes associated with dam planning. The Curitiba Declaration *(see Box 3.2)* went beyond the Manibeli Declaration's almost exclusive focus on the World Bank. It was broader in scope and more confident in its tone. While endorsing the Manibeli Declaration and acknowledging the proposed meeting in Gland, the Curitiba Declaration also called for an independent review of large dams.

Both the Manibeli and Curitiba Declarations reflect the growth and sophistication of transnational alliances. Critics of these developments who argued that participants in transnational alliances came from non-representative processes or that their protests must be articulated only within national boundaries were beginning to acknowledge that these alliances brought substantial research, mobilisation, and understanding to their protests. They represented significant voices, possibly of silent majorities, and could contribute to the definition of global norms. Inevitably, they were also contributing to the process of what constituted multilateralism and how multilateral stakeholder processes should be defined.[14]

Birth of the World Commission on Dams

The role of IUCN as a co-convenor of the Gland workshop proved critical in defining the range of stakeholders who would be represented at the table and subsequently, the legitimacy of the global Commission in the eyes of the anti-dam movement.[15] IUCN staff members were active proponents of diverse participation at Gland. According to a former senior official, IUCN had two basic conditions for getting involved in the workshop.[16] The first was that the Gland meeting would not be about how to build better dams, but that it would be about the planning process for assessing options in water resources and energy management. The second condition was that all parties should be involved "from the most radical activist groups to the most conservative business groups."[17]

IUCN lacked extensive expertise in large dams issues, and its track record in environment and development was not a perfect match, since the thorniest dam struggles related to comprehensive human rights violations: lack of participation, developmental displacement, rehabilitation, and resettlement. However, IUCN's independent status provided the credibility required to bring community-based activists and advocacy NGOs into dialogue with corporate and multilateral representatives. "In order to bring in the most active opponents of dams we needed to have a partner who had their confidence," noted one World Bank official. "The Bank couldn't imagine a serious partnership with IRN."[18]

Meanwhile, the reputation of the World Bank and its continuing importance as an underwriter of private sector investments in developing countries provided the weight required for private sector participation at Gland.[19] Furthermore, the issues highlighted in the OED Review provided sufficient resonance with a subset of businesspeople to bring them to the table. As one representative from the more liberal segment of industry would later note, "Industry wants to learn. The past experience with dams has not always been positive, we had underestimated the technical, economic, environmental and social risks."[20] The overall response to workshop invitations was "extremely positively surprising,"[21] in the words of one organiser.

The Path to Gland

By the mid 1990s, private companies involved in the dams business had begun paying heed to environmental issues to varying degrees—whether motivated by civil society pressure, government regulations, or a larger sense of corporate responsibility. For instance, the International Commission

> Box 3.3
>
> **ICOLD Position Paper on Dams and the Environment (excerpted)**
>
> Attention to the social and environmental aspects of dams and reservoirs must be a dominating concern pervading all our activities in the same way as the concern for safety. We now aim at balancing the need for the development of water resources with the conservation of the environment in a way which will not compromise future generations...
>
> In search of this balance, ICOLD members should be guided by the following aspects of environmental policy:
>
> - The larger the project, the greater the effects on the natural and social environment to be expected, and the wider the scope of the multidisciplinary, holistic studies which they require. Large-scale development demands integrated planning for an entire river basin before the implementation of the first individual project(s). Where river basins are part of more than one country, such planning presupposes international cooperation.
> - Projects must be judged everywhere and without exception by the state-of-the-art of the technologies involved and by current standards of environmental care. The scope for reducing any detrimental impacts on the environment through alternative solutions, project modifications in response to particular needs, or mitigating measures should be thoroughly investigated, evaluated and implemented.
>
> A comprehensive Environmental Impact Assessment, since 1971 mandatory in a growing number of ICOLD member countries, ought to become standard procedure everywhere as part of project conceptualisation, that is well before final design and the start of construction.
>
> Source: International Commission on Large Dams, "Position Paper on Dams and Environment," May 1997.

on Large Dams (ICOLD) developed a *Position Paper on Dams and the Environment* in 1997.[22] ICOLD is a professional association of dam builders from private sector and government and the owner of the largest repository of technical dam-related data in the world. The content of the position paper was fairly general in nature. *(See Box 3.4.)* Members considered it to be a major step for the association.[23]

Meanwhile, a smaller group of private companies and state utilities were beginning to explore ways in which they could improve their environmental performance under the auspices of the International Energy Agency (IEA) Hydropower Agreement, established in 1995.[24] This group sought a clearer standard for accountability that could decrease uncertainty in investments and boost the competitiveness of hydropower vis-à-vis other energy sources. The IEA model acknowledged the need for sound decision-making process, options assessment, and community relations in hydropower development.[25]

An industry representative would later attribute these voluntary initiatives to companies' recognition of the new importance of private sector actors: "Dams used to be government projects. But since the 1990s, the role of the private sector has been expanding. This means new rules of the game, new responsibilities."[26] A subset of progressive executives were beginning to realise that they required a license to operate that went far beyond the requirements of bureaucratic licensing to include broader credibility and consent.

The financial and legal costs of civil society dissent also burdened private sector operations, the way they did the World Bank's. "The whole process of dam development was grinding to a halt, that had just become increasingly evident," noted one observer.[27] Therefore, seeking a consensus with dam opponents on a way forward was central to companies' self interest.

At Gland

It was against this background and the continuing struggles against dams across the world that the Gland meeting took place in April 1997. Convened by the World Bank and the IUCN, the two-day workshop brought together 39 participants representing the diverse interests in the large dams' debate. The workshop's objectives were to review Phase I of the OED study, develop a methodological framework for Phase II, and propose a rigorous and transparent process for defining the scope, objectives, organisation, and financing of follow-up work. The objectives were also to more clearly define the scope of the Phase II study, including basic guidelines for involvement by governments, private sector, and non-governmental organisations as well as public participation, information disclosure, and subsequent dissemination of results. Another objective was to identify follow-up actions necessary, including generally accepted standards for assessing, planning, building, operating, and financing large dams that would reflect lessons learned from experience.

The 39 participants at the workshop came from a range of interest groups involved in the planning, construction, management, and opposition to large dams: professional associations, such as the International Commission on Large Dams; companies, such as Asea Brown Boveri and Harza Engineering; think tanks, such as Tata Energy Research Institute; international non-governmental organisations, such as International Rivers Network; and affected people's movements and groups, such as Narmada Bachao Andolan (Struggle to Save the Narmada River) in India. Senior officials attended from the World Bank and IUCN. Invitations were issued to institutions, which nominated their own representatives to attend.[28] In addition, a few people were invited in their individual capacity.

During the Gland proceedings, the authors of the OED Review acknowledged its limitations.[29] According to a report of the workshop proceedings, "there was widespread recognition that further work was essential and that it would need to be comprehensive in scope, transparent in conduct and defensible in its analyses." [30] NGO representatives at Gland also noted, "World Bank representatives seemed rather self-critical and appeared to respect the strength and the arguments of dam opponents… Some World Bank representatives were also openly critical of the Bank's record on dams, of their poor record of implementing their policies, and of the weakness of these policies." [31]

The explicit participation of national governments in the Gland meeting was relatively modest. There was only one ministry official there: from China, which was undertaking possibly the largest modern engineering feat in the Three Gorges Dam—planned for an estimated capacity of 18.2 million kilowatts, a height of 575 feet, and the displacement of up to 1.9 million people.[32] However, various proxies and quasi-governmental appointees represented government viewpoints. The World Bank's senior staff members were familiar with the views of client governments and industrialised country donors on their board. The Gland meeting included representatives of two state-owned utilities[33] and two river basin authorities.[34] The ability of the WCD to attract the political support of governments would become a serious issue in its later process and would inhibit a more enthusiastic engagement by government agencies and officials with the proceedings and the eventual report. NGOs active in the formative processes felt strongly that while the composition of the Commission had to have senior people with high credibility with governments, integrally involving governments from the inception would have compromised the Commission's ability to produce an unbiased report.[35] These political trade-offs became part of the many lessons that the WCD process has subsequently highlighted.

The gender profile of the Gland meeting participants was predominantly male; this was partially a product of the overwhelmingly male dominance in the dams business, although not characteristic of civil society organisations as a whole. Of the 39 participants (excluding media and observers), only 2 were women: a senior social scientist at the World Bank and a Chinese scientist. Although successful efforts were later made to achieve gender balance on the Commission itself (5 of the 12 Commissioners were women),[36] the Gland meeting foreshadowed the difficulty of integrating women's perspectives in all aspects of the subsequent WCD process.

Very few women participated in the Gland meeting.

The agenda for discussion at Gland indicated the distance that the dam-building establishment had agreed to travel to address the controversy around large dams. Included in the agenda was a comprehensive review of large dams around the world and the decision to define comprehensive standards for the building of large dams. A senior water expert of the World Bank reflected this changed mood when he acknowledged that Bank-funded dams had caused serious adverse impacts on land and people. He said that the independent assessment was an idea whose time had come because the World Bank "didn't have the moral authority to make these judgements."[37] Although the stated intent of the Gland meeting was to work toward a second phase of the OED study, NGOs' calls for an independent review prevailed.

The workshop was the result of exhaustive preparation by the convening organisations. Mindful of the tremendous opportunity the gathering posed, they were determined not to let it slip away. "My major concern was not to have a conflagration," said the facilitator, "but to establish and maintain a conversation among participants."[38] The convenors entered the meeting with a comprehensive set of contingency plans on how to react to debates and a set of scenarios, ranging from minimalist goals of getting people to agree to meet again to a much weightier scope of future co-operation.[39]

The World Commission on Dams Takes Shape

A spirit of constructive debate prevailed in the discussions. The meeting unanimously stressed the urgent need for the second phase of the OED Review to include all large dams and not just those funded by the World Bank. Discussions also underscored the need for a representative team of eminent people to carry out the review independently. Consequently, the participants at Gland agreed to proceed with an independent commission to review the performance of large dams and develop guidelines for the future. It was a breakthrough in civil society's relationship with planners, dam builders, and financiers that the proposed commission's mandate emphasised both the review of the "development effectiveness of large dams" and an assessment of the alternatives. What was remarkable was that an independent process had been agreed upon by a group of stakeholders who had seemingly irreconcilable differences about the value of large dams—differences that had generated some of the most acrimonious conflicts in post-war development history.

Participants at the Gland meeting articulated the terms of reference of the World Commission on Dams as follows:

- "To assess the experience with existing, new, and proposed large dam projects so as to improve existing practices and social and environmental conditions;
- To develop decision-making criteria and policy and regulatory frameworks for assessing alternatives for energy and water resources development;
- To evaluate the development effectiveness of large dams;
- To develop and promote internationally acceptable standards for the planning, assessment, design, construction, operation, and monitoring of large dam projects and, if dams are built, ensure affected peoples are better off;
- To identify the implications for institutional, policy, and financial arrangements so that benefits, costs, and risks are equitably shared at the global, national, and local levels; and,
- To recommend interim modifications—where necessary—of existing policies and guidelines, and promote "best practices."[40]

Participants agreed on the outlines of the institutional structure that the World Commission on Dams would take. The Commission would comprise between five and eight Commissioners, including an "internationally recognised" chairperson. The Commissioners would have "appropriate" expertise and experience and would be widely regarded as having integrity and representing the diversity of perspectives, including affected regions, communities, and private and public sectors. This criterion would prove highly contentious when the time came for the actual selection of Commissioners. Most of them would serve part-time while up to three would serve full-time. The Commissioners would serve in their personal capacities.

The workshop agreed on a tripartite institutional structure for the WCD. Apart from the Commission, there would be a consultative group composed of participants who attended the workshop, plus others invited by the Commission from NGOs, multilateral institutions, governments, and the private sector. This Forum would ensure effective and balanced representation of all stakeholders and key actors. The group would be used as a sounding board for ideas. A third element in the proposed institutional structure was a Secretariat of full-time professionals who would provide support for the Commissioners.

Another important element of the Gland agreement was an outline of the Commission's operational mandate. Its *modus operandi* would consist of study groups, public hearings, commissioned studies, and task forces. Agreement was also reached on an implementation strategy to take effect immediately after the workshop. IUCN's Director General agreed to set up an Interim Working Group (IWG) composed of IUCN and World Bank staff. The IWG would draw on participants in the workshop for advice and support in establishing the Commission. By the end of October, the IWG would establish full terms of reference for the Commission and its advisory Forum, membership of the Commission and the Forum, capabilities and location of the Secretariat, an outline program and budget for the Commission and Secretariat, and a funding strategy for the two-year life of the Commission.

In the weeks that followed the Gland Workshop, the IWG was formed with six representatives from the World Bank and five from the IUCN. In deference to NGO participants in Gland who were uneasy working with the World Bank, it was agreed that IUCN would take the primary managing role. The IWG was given six months to decide on the composition and procedures of the WCD in consultation with members of the gathering at Gland—called the Reference Group (RG).

Conclusions

The conditions for the genesis of the WCD lay in the global and national maturity of the dams debate. A critical mass of local, national, and transnational civil society organisations had emerged to challenge not just the social and cultural injustices and environmental degradation and damage caused by some of the worst large dam projects but also the conceptualisation, planning, and implementation of dam-building projects. Civil society groups, who most desired a change in the status quo, were the primary proponents of an independent commission. Growing evidence of the adverse impacts of large dams as well as the high transaction costs of enduring civil society opposition led the World Bank to question the extent of its future role in supporting the building of these dams. Certain private investors in large dams also sought clearer ground rules for their engagement with host governments and communities in order to free them of damaging charges and smooth their operations. Additionally, environmental issues were also beginning to have greater resonance within the professional dams community. These complex realities and concerns among diverse stakeholders, and the WCD's own evolving commitment to be a democratic multi-stakeholder forum, convinced stakeholders to participate—if they had an equal place at the table.

The acrimony of the dams debate meant that the risks of multi-stakeholder engagement were high, including for those in the World Bank who were attempting to pursue greater participation and openness. All the participants in the Gland meeting had a sense of entering truly untested territory.

The addition of IUCN as a convening body, along with the World Bank, was critical to opening the Gland meeting to participation from other stakeholders. IUCN also helped to make NGOs more confident to attend the meeting. Sustained pressure from civil society organisations and social movements compelled the convenors to both widen participation and influence the content and process of the Gland meeting.

IUCN's role as a convenor was critical to broad stakeholder participation.

Participants unanimously identified the need for an independent and comprehensive review of the development effectiveness of large dam projects—above and beyond the World Bank's own projects—and the generation of guidelines for future water and energy resources development. The suggestion for a WCD emerged as a promising response for all participants.

The setting up of the Commission was a major step forward for multi-stakeholder processes. It was a step with the potential to influence and perhaps even define global norms for the building and functioning of large infrastructure projects and other development processes. The number and nature of variables that led to the formation of the WCD will not be the same for other contentious issue areas. The principle of using transparent and inclusive multi-stakeholder consultations to define a commission is a relevant pre-condition, but the precise process and outcome may vary. The formation of such processes calls for continuing innovation and creativity on the part of all those seeking to democratise global and national policy arenas.

Endnotes

1. One of the many examples is the Almatti Dam in India.
2. For instance, P. McCully, *Silenced Rivers* (London: Zed Books, 1996); E. Goldsmith and N. Hildyard, *The Social and Environmental Effects of Large Dams* (San Francisco: Sierra Club Books, 1984). See also the website of the International Rivers Network, www.irn.org, for dams in Europe and the U.S. slated for decommissioning.
3. Interview with former member of the World Bank Inspection Panel, 16 November 2000. See Sanjeev Khagram, "Toward Democratic Governance for Sustainable Development: Transnational Civil Society Organizing around Big Dams," in *The Third Force: The Rise of Transnational Civil Society*. Ann Florini, ed. (Washington, DC: Carnegie Endowment for International Peace, 2000); Lori Udall "The International Narmada Campaign: A Case of Sustained Advocacy," in *Towards Sustainable Development? Struggling Over India's Narmada River*. William F. Fisher, ed. (New York: ME Sharpe, 1995).
4. McCully, 1996, p. 302.
5. Patrick McCully, "A Critique of 'The World Bank's Experience With Large Dams: A Preliminary Review of Impacts,'" April 1997. Online at: www.irn.org/programs/finance/critique.shtml (28 September 2001).
6. Interview with OED official, 7 September 2000.
7. World Bank Operations Evaluation Department, *World Bank Lending for Large Dams: A Preliminary Review of Impacts*, OED Précis, September 1996.
8. Interview with OED official, 7 September 2000.
9. World Bank Operations Evaluation Department, 1996.
10. Interview with World Bank official, April 2000.
11. World Bank/IUCN, *Large Dams: Learning from the Past, Looking to the Future* (Gland: IUCN, 1997).
12. McCully, 1997.
13. IRN press release, "World Bank Dam Evaluation 'Seriously Deficient': NGOs Demand Independent Review and Moratorium on World Bank Support of Large Dams," 7 April 1997. Online at: www.irn.org/programs/finance/pr970407.html (28 September 2001)
14. See Chapter 2 for a more detailed assessment of multilateralism and multi-stakeholder processes.
15. Interviews with involved IUCN and World Bank staff, April and July 2000.
16. Interview with former senior official of IUCN, 5 September 2000.
17. Interview with senior IUCN official, 5 September 2000.
18. Interview with OED official, 7 September 2000.
19. A senior Bank official speculated that "without the Bank a lot of these guys wouldn't be in... Part of it is financing, part is reputational." Interview with senior World Bank official, June 2000.
20. Interview with industry representative on the Forum, April 2000.
21. Interview with Gland workshop organiser, 1 February 2001.
22. For the full text of the Position Paper, see: http://genepi.louis-jean.com/cigb/chartean.html.
23. Interview with former ICOLD president, December 1999.
24. IEA Hydropower Agreement website, www.ieahydro.org (28 September 2000). The ten member countries of the IEA Hydropower Agreement referenced are Canada, China, Finland, France, Japan, Norway, Spain, Sweden, the United Kingdom, and the United States.
25. Interview with IEA co-founder, April 2000.
26. Comments of an industry representative during an industry presentation to World Commission on Dams Forum members, 4 April 2000.
27. Interview with Gland workshop organiser, 1 February 2001.
28. Interview with Gland workshop organiser, 1 February 2001.
29. World Bank/IUCN, 1997, p. 9.
30. World Bank/IUCN, 1997, p. 9.
31. P. McCully, P. Bosshard, and S. Dharmadhikary, *An NGO Report on the April 1997 World Bank-IUCN Dams Workshop and on the Proposal for an Independent International Dam Review Commission*, 29 May 1997 (mimeo).
32. These widely publicised figures are confirmed in the WCD's country review of China's dams. R. Fuggle; W.T. Smith; Hydroconsult Canada Inc.; and Agrodev Canada Inc. 2000. "Large Dams in Water and Energy Resource Development in the People's Republic of China (PRC)," country review paper prepared as an input to the World Commission on Dams, Cape Town. Estimates for the number of people to be displaced by the Three Gorges Project vary. The figure of 1.9 million is cited on International Rivers Network's website, http://irn.org/programs/threeg/.
33. Electricité de France and ISAGEN-Colombia.
34. Volta River Authority and Lesotho Highlands Development Project.
35. Interviews with Forum member, September and November 2000; e-mail communication with Forum member, January 2001.
36. This was, unfortunately, not the experience with the Secretariat, which had few women.
37. Interview with World Bank Senior Advisor, June 2000.
38. Interview with Gland workshop organiser, 1 February 2001.
39. Interview with Gland workshop organiser, 1 February 2001.
40. World Bank/IUCN, 1997, p. 10.

Chapter 4

From Gland to Cape Town: The Making of the WCD

The question of who participates and how is at the heart of any multi-stakeholder process. The WCD's ability to create and maintain legitimacy depended on whether different stakeholder groups felt adequately represented in its process. In this chapter, we examine the role of the Commission, Secretariat, and advisory Forum and how they came to incorporate a range of interests. The acceptability of the Commission, Secretariat, and Forum to different groups holds implications for the design of future commissions. In addition to analysing representation, we examine the funding base of the WCD and the design of the workplan, which reflect concerns about independence and inclusiveness in the design of the WCD.

Representation of Viewpoints on the Commission

Multi-stakeholder processes (MSPs) raise complex questions of representation, and the WCD was no exception. Within the established framework of democratic government, the ballot box is the accepted test of representation. For MSPs, however, representation is a far more murky construct. Members are not elected and their constituencies are not clearly defined. In the case of the WCD, where Commissioners served in their personal capacities, there were no commonly accepted mechanisms by which they were held accountable to stakeholder groups. Yet, ensuring that different stakeholder views are represented is central to MSPs, particularly those created to provide advice on contentious issues, such as large dams. Thus, a key ingredient in the recipe proposed by the Gland workshop was appointment of Commissioners who were "representative of the diversity of perspectives."[1] In this section, we explore how the WCD sought to tackle the question of representation to ensure that participation by various stakeholders strengthened, rather than diluted, the legitimacy of the process. Although the primary focus in this chapter is on the selection of Commissioners, we also examine how the formation of the Secretariat and the Forum are relevant to issues of representation.

A Protracted Struggle over Commissioner Selection

In the months immediately following the Gland workshop, the institutional apparatus of the Commission was not yet in place. During this time, stakeholder groups pressed the World Bank and IUCN for a role in forming the Commission. An influential report on institutional design penned by the facilitator at Gland, Professor Anthony Dorcey, noted the importance of expanding the legitimacy of the Commission beyond that provided by the authority of the two convening organisations, the World Bank and IUCN. *(See Box 4.1.)* At stake was the perceived independence of the Commission. The selection of Commissioners rapidly became a flashpoint for dispute.

The implementation strategy agreed in Gland called for the Interim Working Group (IWG), composed of a handful of senior World Bank and IUCN staff, to devise the terms of reference for the Commission and select Commissioners. It was to draw on the Gland participants (Reference Group) for "advice and support."[2] This vague formulation quickly led to an argument over the degree to which Reference Group members would have a say in the Commission's structure and composition. As one participant in the Gland meeting suggested, "...people left the meeting with a different sense of what had been agreed to! In some ways they wanted to leave before it all fell apart."[3] Hence, the

Box 4.1

Foreshadowing challenges and pitfalls: recommendations of the Dorcey Report

Professor Anthony Dorcey, the facilitator at the Gland meeting, addressed design questions in a report written at the request of the Interim Working Group. The Dorcey report was influential in shaping key dimensions of the Commission, and prescient in pointing out potential challenges and pitfalls along the road.

Structure

1: Consult chairpersons and secretaries-general of selected previous international commissions.

2: Use the term "commission" and add appropriate qualifiers.

3: Make clear the commission's nature and purpose in the terms of reference and adopted mandate.

4: Be explicit about how the Commission's design will contribute to immediate and longer term improvements in decision making on large dams.

5: Commission design should reflect criteria of inclusiveness, transparency, consensus, effectiveness and fairness, while minimising the trade-offs necessitated by constraints of time and resources.

Initiation and Midwife Organisations

6: Be explicit about the intended independence of the commission and its implementation.

7: Discussions on the terms of reference and their implementation should precede their formal adoption and appointment of the chair.

8: Increase the perceived legitimacy of the commission and facilitate its access to information and resources by going beyond the World Bank and IUCN to obtain formal endorsement from other organisations.

9: Discuss the proposed commission with representatives of the key stakeholder organisations whose interests have been identified as not being adequately reflected in the Reference Group.

Characteristics

10: Make clear that it is an "advisory" rather than an "investigatory" commission.

11: Issue a statement of the commission's adopted mandate, strategy, workplan and expected products as soon as possible after it is established.

12: Make clear that while the immediate goal of the commission focuses on building consensus among the commissioners on conclusions and recommendations, the longer term goal is to provide a foundation for building consensus in large dam decision making processes.

13: Exploit new computer-based technologies and the world wide web to their fullest advantage. But take great care to ensure that they are used as complements to the range of other well-proven communications techniques and that steps are taken to minimise the extent to which stakeholders are disadvantaged by their lack of access to new technologies.

14: Utilise case study assessments of experience with a representative set of large dam projects and, wherever possible, include multi-stakeholder processes.

Personnel

15: Agree on a final set of criteria and the weightings for each of them before proceeding to select from among the nominations for the chair.

16: Agree on a recommendation for appointment to the position of chairperson that would go to the chief executives of the formally sponsoring organisations.

17: All commissioners should be able to commit a comparable amount of time – at least 40 days per year.

18: All commissioners should agree to serve in their personal capacity.

19: Agree on recommendations to the chairperson on the criteria to be used in selecting commissioners.

20: Make recommendations to the chairperson on criteria to be considered in selecting the secretary-general and assist in the search for potential candidates to facilitate an early decision.

Source: Anthony Dorcey, Institutional Design and Operational Modalities for the Proposed Large Dams Commission, Stockholm Draft, 6 August 1997 (mimeo).

stage was set early for a tussle over authority between the IWG and the Reference Group.

The mandate from Gland was to select a slate of Commissioners with "appropriate expertise and experience…widely regarded as having integrity and being objective, independent, and representative of the diversity of perspectives including affected regions, communities, and private and public sectors."[4] The challenge lay in determining the appropriate range and mix of perspectives on dams, and most importantly, deciding who would make this determination. The ensuing process of Commissioner selection, summarised below, was protracted, bitter, and partisan.

Selection of the Chairperson indicated the lenses through which different groups would view the battles ahead. The IWG selected Professor Kader Asmal, then the Minister of Water Affairs and

Forestry in South Africa, to chair the WCD. Each side in the debate saw signs of hope in his selection. As a government minister from the South, Prof. Asmal could reasonably be seen as credible by governments, particularly in Southern dam-building nations. Moreover, the legitimacy conferred on him as an elected politician was of great importance, particularly to governments.[5] Private sector interests supported his candidacy based on his status as a minister in a Southern country with a track record of concern for economic development. They noted his past approval of a large dam in the Lesotho Highlands Water Project complex in South Africa. As one Forum member from the private sector noted, a minister could be counted on to uphold sovereign rights to decision-making over national resources, and not to abdicate authority to non-elected stakeholders.[6] Indeed, Prof. Asmal's personal credibility was the basis on which this observer persuaded his board of directors to participate in the WCD.

Civil society groups were initially wary of Prof. Asmal's selection. In particular, they were concerned about his past support for the Lesotho water project, but were heartened by his distinguished background as a human rights and anti-apartheid activist. Ultimately, they decided to support him as Chairperson, but subject to a tacit understanding that the position of Vice Chairperson would be filled by someone who had their complete trust.[7] The result was a Chairperson who was endorsed by all the stakeholders, albeit with different degrees of confidence. Subsequent events were to bear out the importance of a strong and credible Chairperson to the ultimate success of the Commission.

Selection of the Chairperson was as far as the unanimity of the selection process extended. In particular, the process was marked by contention between the IWG and NGO stakeholders on the Reference Group. NGO and social movement participants rejected the IWG's first list of 10 candidate Commissioners as "an insult." In their eyes, the list lacked direct representatives of affected communities and active anti-dam campaigners. By contrast, they argued that several representatives of the dam construction industry, dam promotion organisations, and technical experts, all of whom could be seen as dam supporters, were nominated to the Commission.[8]

The back-and-forth over proposed lists between October 1997 and January 1998 was characterised by considerable ill-will and hostile exchanges. NGO and social movement representatives on the Reference Group threatened more than once to withdraw from the process entirely. They suggested that the attempt by the IWG and the Chairperson to retain complete authority over the Commissioner selection process would "flagrantly breach the consensus spirit of Gland."[9] In response, the IWG and Prof. Asmal, who was now part of the process, threatened to pursue the process even if it was "without the complete consensus we had hoped for in Gland."[10] In hindsight, the IWG found that the Reference Group wanted more ongoing input than they at first realised. They were operating on the assumption that they had a mandate to move forward and were surprised to be challenged.[11]

Commissioners were to represent a diversity of perspectives.

A compromise was reached only after Professor Asmal threatened to resign if the group could not reach consensus. The compromise provided for an "expanded IWG" with a small number of interest group representatives from the Gland meeting, in addition to the World Bank and IUCN representatives. Thus, after much contention, a wider range of stakeholders won a say in Commissioner selection. In late January 1998, some 3 months late, this body was able to reach consensus on a list of 12 Commissioners. *(See Box 4.2.)* The main change from earlier lists was the addition of Medha Patkar, a founder of the Narmada Bachao Andolan (Struggle to Save the Narmada River).

A Commission Based on Stakeholder Interests

As a result of the difficult negotiations between the World Bank/IUCN Interim Working Group and participants from the Gland meeting, the Commission turned out to be characterised by representation of stakeholder interests, rather than purely by the eminence of its members. In this regard, the WCD was a significant departure from past commissions. The Commissioners of the WCD were active practitioners in international networks, which included NGO, social movement, and business networks, as well as government agencies.

Box 4.2
The Commissioners

Kader Asmal
WCD Chairperson
Ministry of Water Affairs and Forestry, South Africa

Lakshmi Chand Jain
WCD Vice-Chairperson
High Commissioner to South Africa, India

Judy Henderson
Oxfam International, Australia

Göran Lindahl
Asea Brown Boveri Ltd., Sweden

Thayer Scudder
California Institute of Technology, United States

Joji Cariño
Tebtebba Foundation, Philippines

Donald Blackmore
Murray-Darling Basin Commission, Australia

Medha Patkar
Struggle to Save the Narmada River, India

José Goldemberg
University of São Paolo, Brazil

Deborah Moore
Environmental Defense, United States

Shen Guoyi[a]
Ministry of Water Resources, China

Jan Veltrop[b]
Honorary President, International Commission on Large Dams, United States

Achim Steiner[c]
WCD Secretary-General
Germany

[a] Resigned, early 2000

[b] Joined Commission in September 1998 to replace Wolfgang Pircher, the original nominee

[c] Initially an ex-officio Commissioner

Note: Affiliations as of May 1998
Source: World Commission on Dams, Interim Report, July 1999.

The emphasis on representation went hand in hand with greater stakeholder involvement in Commissioner selection, because only stakeholders, not the IWG, could determine who was representative of their interests.

This is not to suggest that it was easy to define stakeholder groups and establish who could represent them. The stakeholders who selected the Commissioners were an ad hoc group of participants from Gland chosen for their diverse political views and prominence in the dams debate. They sought balance on the Commission between the North and South and diversity of experience in government, private sector, community organising, environmental management, academic disciplines, and other related issues.[12] The greatest benefit from their participation was that the stakeholders could determine the political acceptability of the Commission as a whole. The WCD could later claim legitimacy based on its incorporation of the entire spectrum of views and perspectives in the dams debate. *(See Box 4.3 for one example.)*

The emphasis on representation, rather than eminence, would have far-reaching implications for Commissioners' relationships with interest groups in the debate and on stakeholders' subsequent support for the Commission's work.[13] On the one hand, Commissioners were explicitly invited to join the Commission as individuals and not as formal representatives of an organisation. "When I was invited to participate, my chief criterion was independence," said one Commissioner. "If you're beholden, you're stuck."[14] Such independence was essential if Commissioners were to develop shared understanding with each other and move toward a consensus.

On the other hand, Commissioners' individual legitimacy stemmed from their prominence in international networks and their unspoken claim to represent certain viewpoints. As a result, they faced a perpetual tension between forging consensus among themselves, which required flexibility, and maintaining the trust of constituents, which required a measure of rigidity. When asked whether she was a representative of any group, one Commissioner responded: "Yes and no. It's very complex. I have a responsibility to represent a point of view of a particular constituency. It was in the mandate that we do not represent our organisations for the reason that I don't go back to them on everything I have to say. They have expressed faith in my nomination."[15] We return to the subject of representation and how Commissioners balanced competing pressures in Chapter 7.

The struggle by NGOs for participation in decision-making, rather than consultation alone, set a benchmark for stakeholder engagement throughout the process, but only after it had created an initial climate of distrust.[16] In order to rebuild trust, the IWG wrote to Reference Group members promising to implement the Gland commitment

Box 4.3

Defining allies-the perspective of the Movimento dos Antingidos por Barragens *(Movement of Dam-affected People, Brazil)*

The views of the Movement of Dam-affected People (MAB), Brazil, on the composition of the WCD illustrate how stakeholders saw their perspectives represented on the Commission. MAB's leadership believed that the majority of WCD Commissioners were pro-dam in orientation. The Commissioner from Brazil was a former government official whom they considered to be a historical enemy of their movement.

Nonetheless, MAB's view of the Commission was tempered by the inclusion of Medha Patkar, the anti-dam activist, and the broader involvement of the Struggle to Save the Narmada River in the process. MAB saw Ms. Patkar as highly qualified and capable of playing a significant role in the process.

> "During the choice of Commissioners, MAB almost left the process. We had many problems. But the inclusion of Medha Patkar made us continue, because if the Save the Narmada was participating we had to participate too."[a]
>
> "I am sure that there were more pro-dam representatives than anti-dam ones. But I think that our Commissioners were so well qualified that despite the minority in numbers we could easily present our ideas and convictions."[b]

[a]Interview with NGO activist, MAB's IV National Congress, November 1999.

[b]Interview with a leader from MAB, MAB's IV National Congress, November 1999.

Source: Based on Flávia Braga Vieira, "Brazil's Dam-Affected People Movement and the World Commission on Dams." Background paper for the WCD assessment, January 2001.

to a fair and transparent process not only in the preparation phase but also throughout the life of the Commission.[17]

At least two lessons about developing a multi-stakeholder process emerge from this brief tour of the WCD's early history. First, decision-making authority over commissioner selection must be clear. The ambiguous authority given to the IWG, tempered by the Reference Group's consulting mandate and Prof. Dorcey's calls for greater stakeholder engagement, led to strains on the emerging dialogue between stakeholders. Second, if the legitimacy of a commission is based in part upon perceptions of commissioners as representative of broader interests, as is the case with the WCD, then the process of selection must take stakeholder groups into confidence. The participation of a small group of stakeholders from Gland helped gauge the legitimacy of individual Commissioners in professional and interest-based networks, and the political acceptability of the Commission as a whole.

Reflections on the Composition of the Commission

One consequence of the Commission's composition along sectoral and disciplinary lines was that regional and country representation were not perfectly balanced. The Commission included nationals from the world's four largest dam-building countries: China, India, Brazil, and the United States. It was equally balanced between North and South, with both the Chairperson and Vice-Chairperson coming from the South, where the majority of future dams were forecast to be built. Both of these aspects boded well for the Commission's success. However, the fact that the Commissioners came from a total of seven countries and included two Indians, three Americans, and two Australians perplexed stakeholders in other major dam-building regions, such as Latin America, East and West Africa, and the Middle East.[18] Although multi-stakeholder processes are composed to reflect diverse political and sectoral views, the WCD experience suggests that regional representation still matters.

The Commissioners were active practitioners in international networks.

Secondly, the selection process sought to balance gender diversity and was quite successful in doing so. Of the 13 original Commissioners (including the Secretary-General), 5 were women. This level of participation by women was quite notable by historical standards. The caveat is that all but one of these five women came from civil society backgrounds. With the exception of Shen Guoyi from the Government of China, the private sector and government participants were all male. This formulation was to affect, in subtle ways, the dynamics within the Commission as, by and large, the male participants came from positions of significantly greater power than did the female participants.[19]

The Effect of Stakeholder Mobilisation on Representation

The degree to which stakeholder groups approved of the Commission's composition—and felt their views were represented—depended partly on how organised they were to participate in the early process. They were not all equally engaged in advocacy.

NGOs and social movements were extremely organised and, to an external audience, appeared united in their engagement in the Commissioner selection process.[20] As a result, they nominated Commissioners who enjoyed wide credibility within civil society—with backgrounds that spanned the interests of development, environment, indigenous peoples, and resettled populations. Although civil society groups continued to have reservations about other Commissioners and there was some internal dissension, most groups agreed to continued participation and engagement in the WCD.

Private sector engagement with the IWG lacks a paper trail and so is harder to reconstruct than that of the NGOs and movement groups. An official from the Harza Engineering Company and a past president of the International Commission on Large Dams (ICOLD) provided a dams industry perspective to the Commissioner selection process as part of the expanded Interim Working Group. However, there is little evidence of broad private sector interest in and engagement with the WCD at this early stage. Certainly, the dams industry did not have networks in place to co-ordinate common interests the way that anti-dam and social justice groups did. Many companies that had a stake in the WCD were either unaware of the Commission, or did not appreciate its potential significance for their operations. For example, a large utility sent a representative to the Gland workshop, but then played no role in Commissioner selection or, indeed, in any aspect of the WCD until very late in the process.[21] Even among companies that appreciated the relevance of the WCD, private sector participants suggested that competition among firms placed barriers to working in coalitions.[22] As a result, private companies, utilities, and industry associations were less enthusiastic about the Commissioners from industry backgrounds, compared to NGOs and social movements' support for Commissioners from their backgrounds.

Furthermore, when the IWG proposed to include an office-bearer from ICOLD on the Commission, there was considerable contention over which person was most suitable. The original nominee, a former president of ICOLD, attended the first meeting of the Commission then withdrew.[23] Eventually, Dr. Jan Veltrop, another former president, was nominated and accepted. ICOLD members noted that there was vigorous debate behind closed doors at ICOLD on the suitability of Dr. Veltrop's nomination, implying that he did not enjoy the full trust of the membership.[24] In short, the private sector was complacent in the early days of the process, did not sufficiently appreciate the possible impacts of the WCD, and was relatively unprepared and fractious in its demands for representation.

There was little private sector interest in the WCD in the early stages.

Governments were perhaps the hardest group to accommodate within a representational model. Although governments certainly jostle for position and representation when it comes to inter-governmental bodies, multi-stakeholder processes have not, as yet, commanded that same attention. The WCD was no exception. It was up to the World Bank and IUCN and other members of the expanded Interim Working Group to reach out to governments and inform them about the new opportunity for dialogue and shared learning. Standing as they did outside the United Nations or other formal inter-governmental bodies, the onus was on the convenors to ensure that the WCD would be credible to governments.

Ultimately, the Commission departed from the model of past global commissions that have almost entirely comprised eminent retirees from government; however, its members' collective government experience was significant. Three Commissioners were active in government service during their time on the Commission: Professor Asmal of South Africa; L.C. Jain of India; and Shen Guoyi of China. A former government official of another huge dam-building nation, José Goldemburg of Brazil, was appointed to the Commission. Several other Commissioners were either former government officials, or officials of parastatal organisations, such as Don Blackmore, the chief executive of the world's largest river basin authority. The Chairperson frequently pointed out that he and the Vice-Chairperson were from the South, to enhance the Commission's legitimacy with Southern governments.

This profile of governmental experience, while satisfactory to many stakeholders from the broad middle of the dams debate, proved unsatisfactory to some governments. The Commissioner from China's Ministry of Water Resources withdrew midway through the process, apparently because her government was becoming uneasy with the range of views on the Commission. *(See Chapter 7.)* This loss undoubtedly bruised the overall Commission's credibility.

Meanwhile, the Government of India's Central Water Committee perceived both Commissioners from India as being anti-dam and complained that it should have had a greater role in the selection process.[25] Two Indian nationals sat on the Commission: L.C. Jain and Medha Patkar. Ms. Patkar was the leader of the Struggle to Save the Narmada River and National Convenor of the National Alliance of People's Movements. Dr. Jain was India's High Commissioner to South Africa and a former member of India's National Planning Commission and a special committee to review aspects of the Sardar Sarovar Project. Both were critical of dams, Patkar famously so. Although the overall Commission included engineers, investors, and pro-dam planners, it was difficult for officials in India's Central Water Committee to see beyond the apparent bias of the two Indian members. The Indian case illustrates the juggling act that is required to make sure that balance is achieved across a range of parameters. It also highlights that for a global commission a global balance is necessary, but from the viewpoint of nation states, the more important lens is a national one.

The Indian government's preference for greater governmental participation in the WCD selection process and subsequent events illustrates a political trade-off faced by the WCD's convenors. Indian water officials' disapproval of the unfolding WCD process and eventual outcome was partly because of their lack of involvement in the formative process. However, had governments been more involved in the preparatory and subsequent phases of the WCD, the direction of the process would have been different and the political space for inclusion of NGOs and social movements would have been destroyed. An influential NGO activist has stated that the anti-dam movement's involvement was contingent upon weak governmental involvement, because it was governments' water and energy policies that protest movements fundamentally opposed.[26]

The Government of India perceived both Commissioners from India as being anti-dam.

Similarly, the options for greater World Bank influence over the Commission posed a political trade-off. The IWG never considered nominating an official of the World Bank to serve on the Commission because the WCD was rooted in the call by civil society for a truly independent review of large dams. Had a World Bank official served on the Commission (or had the process been tied to the World Bank in other ways), the major anti-dam NGOs and social movements would have left the process and barraged it with negative publicity. They considered the World Bank to be too deeply invested in large dams technology to be capable of providing an independent assessment of dams' benefits and costs.

In future multi-stakeholder processes, questions will inevitably arise about whether powerful actors, such as the World Bank, should have a direct voice in the negotiations, given the tenuous yet possible eventuality that such involvement could increase the institution's buy-in to the results. As with government officials, it is unclear whether and how a World Bank representative could comfortably manage the ambiguous representation—as an individual but also as an implied institutional spokesperson—required to participate in the Commission. Just as with governments, the individual chosen to serve from the World Bank—and his or her clout within the institution—would make a tremendous difference to the outcome of the multi-stakeholder process and institutional acceptance of the results.

In sum, Commissioner selection was a highly complex balancing act. In the case of the WCD, NGOs were the most vigorous in representing their interests. In many respects, the creation of political space for their inclusion depended upon pro-dam interests being less organised and NGOs being able to claim the space for negotiation. In other words, the convenors of future processes must be aware that for contentious development issues such as dams, inclusion of diverse voices is possible but political trade-offs inevitably arise. The WCD experience suggests that future multi-stakeholder commissions should respond to the challenges of preparation and formation with comprehensive mapping exercises of concerned stakeholders and connected selection processes among stakeholder groups, rather than exclusion of any relevant constituency.

The Significance of the Secretariat's Role

As Commissioners were the public and political face of the WCD, the process and outcome of Commissioner selection were critical to representation and, therefore, to the legitimacy of the WCD. By contrast, it is not immediately obvious why the composition of the Secretariat, designed as an administrative body, should also have played a part in stakeholder perceptions of the WCD's legitimacy. And yet, it clearly did. Private sector actors and NGOs concluded that the Secretariat was a critical element of the Commission's structure, and that the composition of the Secretariat shaped outcomes. As we will discuss in Chapter 5, in operationalising the WCD's work plan, the Secretariat's role extended into framing decisions for the Commission in important, if understated, ways. As a result, concerns over the representation of interests became an issue within the Secretariat, just as much as within the Commission.

In processes of this nature, the importance of the Secretariat depends heavily on how Commissioners choose to structure their work. In the early days of the WCD, some Commissioners proposed that each of them conduct an independent work programme using WCD resources, and convene occasionally to knit these disparate elements into a cogent final report. This model would have required only a minimal Secretariat, one which would have had less influence in shaping outcomes and, therefore, fewer implications for the WCD's overall legitimacy.[27] However, this model was abandoned as the Chairperson argued that only through a collective, unified approach would the Commission have sufficient legitimacy to transcend the partisanship of the dams debate. The Commissioners adopted a single plan that they would supervise together. One significant outcome of this decision was the need for a substantial Secretariat staff to undertake a more comprehensive work programme sanctioned by the WCD. However, this decision put much of the burden of maintaining credibility on to the Secretariat.

Stakeholders concluded that the Secretariat's composition shaped the outputs of the Commission.

Achim Steiner, Secretary-General, in consultation with the Chairperson, shouldered much of this burden. As a Deputy Director with IUCN, Mr. Steiner had been centrally involved in organising the Gland meeting. Further, for several months he guided the early establishment phase of the IWG as its Interim Co-ordinator, until he left to take up another position. When the time came to select a full-time Secretary-General, Mr. Steiner was summoned back to the WCD. His appointment apparently owed much to the relationship he enjoyed with Prof. Asmal, whose trust and confidence he had gained during the initial establishment phase. Indeed, during the course of the WCD, the close working relationship between Prof. Asmal and Mr. Steiner was repeatedly mentioned as an important glue for the process as a whole.[28]

The Selection of Secretariat Staff

Given the Secretariat's important role in running the work programme, stakeholders watched closely to see whether their interests were reflected in the its make-up. Recognising the significance of the Secretariat's composition in gaining stakeholder groups' confidence, the Chairperson and Secretary-General sought staff with wide-ranging views in the dams debate and a diversity of sectoral expertise. As with the

Commission, the effort was to ensure that the Secretariat, as a whole, demonstrated balance. Thus, some staff brought a history of engagement with civil society groups, while others had worked on electricity and irrigation infrastructure projects. To a limited extent, this composition helped pair Commissioners with particular staff with whom they shared a perspective and background, and facilitated communication between the Commission and the Secretariat.

As with Commissioner selection, NGOs and social movements (including Commissioners from those backgrounds) were most active in identifying candidates for the Secretariat and encouraging them to apply. These efforts helped ensure that civil society viewpoints were represented within the Secretariat. Private sector actors and dams associations also played some role in proposing candidates, but did not do so until quite late in the selection process.

Stakeholder groups on the Forum remained uncomfortable with the Secretariat's composition for much of the process, although the issue did not lead any of them to withdraw. Civil society groups charged that the Secretariat staff tended to be drawn from a mainstream perspective associated with international development institutions, such as the World Bank.[29] Critics argued that development agency experience fosters a mind-set that emphasises technocratic approaches over the more political views put forward by peoples' movements and some NGOs. Meanwhile, industry stakeholders noted that several Secretariat staff had previously worked at IUCN, one of the WCD's convening institutions.

Charges of mainstream bias and IUCN tilt had some factual basis: four staff members brought considerable experience working in bilateral or multilateral development agencies. Three had considerable work experience with IUCN. Three more were drawn from academic or research organisations, two of whom had some experience consulting for development agencies. Two brought considerable media experience to the WCD.

Although it is a relatively simple matter to look at patterns in experience, it is much harder to assess what these patterns mean. Some of the effects of past experience, such as the ways in which problems are framed, are difficult to analyse empirically. It is easier to study the extent to which past experience determined the networks that Secretariat members brought to their jobs. These networks were significant to the work of the WCD. Interviews suggest that Secretariat staff relied heavily on their past contacts to identify consultants to carry out the work programme and participants in the consultations organised by the WCD.[30] For example, the case study consultants in Pakistan were selected based on past work conducted by three senior advisors in Pakistan. This evidence of the importance of social networks suggests that explicit attention to the breadth of past experience within the Secretariat might have made for a more inclusive process.

If we accept that the personal experience of Secretariat staff inevitably affects the work of the Commission in explicit and subtle ways, then it is also relevant to examine the composition of the Secretariat along other parameters. Examined for gender balance, the Secretariat was unbalanced, with only two women (one of whom left during the process) among ten senior advisers. However, the Secretariat did also rely heavily on highly qualified interns, of which three-fourths were women.[31] Thus, women were well represented at junior levels, but under-represented at senior levels.

The geographic balance across regions was also uneven. Of the ten senior advisors, three were from Europe, three from North America, two from South Asia, and one each from Latin America and Africa. Admittedly, these categorisations are crude, since most of the Secretariat staff had varied backgrounds and experience in several regions. In particular, all of the Europeans and North Americans also had considerable experience in Southern countries. In addition, the Secretariat staff explicitly sought out interns from various areas, particularly from those regions where the WCD would host regional consultations.[32]

In sum, the composition of the Secretariat reflected the existing bias in development circles toward development agency experience, a preponderance of men over women, and of Northern rather than Southern origins. Some argue that the demographic characteristics of the Secretariat only reflected the constraints of global society in terms of the requisite skills, training, experience, and availability of personnel. However, the legitimacy of a body explicitly committed to inclusiveness, as the WCD Secretariat was, would have been more easily defended had it successfully found ways to overcome these limitations.

Establishment of the Forum

The participants at the Gland workshop called for roles for diverse stakeholders in the Commission's structure. In addition to the Commission and Secretariat, they envisioned a consultative group of stakeholders comprised of their own number, supplemented by others, to be used as a sounding board for the Commission's ideas.[33] This vision was realised in the WCD Forum. The intent was by no means for the Commission to be accountable to the Forum on an ongoing basis. Instead, the Forum was an institutionalised means for the Commissioners, assisted by the Secretariat, to receive feedback on their ongoing work. The Forum would ensure that the wider body of stakeholders remained engaged in the process throughout its life and that the Commission did not drift too far from the wider range of opinions in the debate. Finally, the Forum, as the group of stakeholders most closely concerned with the outcome of the WCD, was intended to publicise the work of the WCD and build ownership for the final product. In the eloquent words of Kader Asmal, if the World Bank and IUCN were the midwives, the Forum was the family into whose hands the fledgling WCD report would be delivered.[34]

The participants in the Gland workshop formed the core of the Forum. It was subsequently expanded to its final size of 68 organisations.[35] As with the other organs of the WCD, the initiators sought a balanced representation across various stakeholder groups. Participation in the Forum was by invitation only. The Commission decided upon new members with the help of the Secretariat. Participation was divided into 10 different group types *(see Figure 4.1)*—a classification that brought a balance of views to the Forum as a whole and that provides a possible model for future processes.

The definition of stakeholder groups had considerable implications for the relative representation of the Forum. For example, had the WCD been explicitly conceived of as a "trilateral network"—a simplistic formulation that has gained popularity in recent years—there would potentially have been considerable pressure to allocate Forum representation equally among governments, civil society organisations, and the private sector. Instead, the WCD chose to adopt a more fine-grained approach based on a close examination of the dams arena. For example, it included separate categories for NGOs and dam-affected peoples to reflect the different perspectives, approaches, and concerns of these two groups. In another example, it divided industry groups into river basin authorities, utilities, and private sector firms. This was entirely appropriate and continued the more sophisticated mapping of stakeholder categories evidenced in Agenda 21 and its follow-up processes. As one commentator wrote of the approach in Agenda 21, "How can you put together NGOs, women, trade unions, scientists, and local government, to mention a few, in one grouping called civil society?"[36]

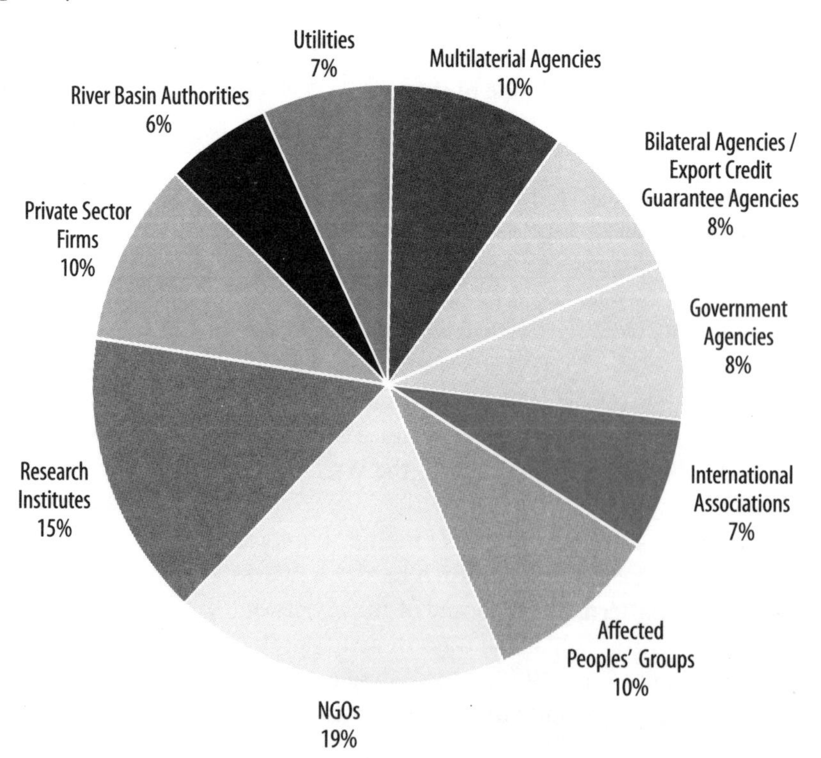

Figure 4.1 Forum composition

Source: WCD website, www.dams.orgabout/forum_list.htm (28 September 2001).

The WCD chose a fine-grained approach to stakeholder representation on the Forum.

Inclusiveness was but one concern in designing the Forum; the other was balance. The different interest groups on the Forum, interestingly, all considered it unbalanced in some way. Industry representatives found the Forum biased towards NGOs and felt that they would not "get their way" until the balance shifted.[37] When community groups and anti-dam NGOs tallied up numbers, they counted many more Forum members from the development establishment, which they considered pro-dam, than from project-affected communities. In the politically charged atmosphere around dams, no single formulation could satisfy all constituencies. The Commission made a broad effort to include multiple voices in approximately level proportions. Its ability to hold the Forum together through the process suggests that the general balance of the WCD's Forum was right.

The Participation of Women

One shortcoming of the WCD's stakeholder formulation is that participants who brought a gender perspective were not explicitly included as a separate stakeholder group. The Commission did acknowledge that there should be a slot for women's issues. The international women's network DAWN (Development Alternatives for Women in a New Era), which has a rotating secretariat among Southern countries, filled this slot.

Otherwise, gender concerns were captured only incidentally by individuals nominated to the other stakeholder categories. In most cases, the institutions chosen for Forum membership nominated male representatives. (The same phenomenon occurred at the Gland meeting, when IUCN and the World Bank invited institutions, not individuals, to participate and the overwhelming majority of representatives were men.) Although it is true that women seldom form a significant political constituency in dams debates, they do form one of the most important stakeholder groups in overall water use and management. Indeed, the Dublin Principles, agreed by governmental representatives in 1992 in the run-up to the United Nations Conference on Environment and Development, recognised, as one of four overarching principles, that "[the] pivotal role of women as providers and users of water and guardians of the living environment has seldom been reflected in institutional arrangements for the development and management of water."[38] The call for women's integral involvement in water management was one of four pillars in the final conference declaration.

The space given to women's voices on the Commission itself was extremely important—and to some degree corrected for the Forum's male bias—in terms of women's influences on the process' outcome and overall public perceptions of gender bias. However, the Forum's tilt nonetheless established it as a place where women had a hard time making themselves heard. Women participants—who were overwhelmingly from NGOs—reported how facilitators and rapporteurs of the Forum meetings disregarded their comments. A member of the sole women's group on the Forum was visibly distressed by the lack of acknowledgement of women's issues in the discussion. An observing female journalist from India wrote an article following the final Forum meeting that highlighted the preponderance of men at WCD meetings.[39]

The Forum lacked the voices of women and gender advocates.

For a global process that was committed to good governance, there is enough evidence to suggest that gender bias in the WCD Forum led women to believe it was not a legitimate space for equal dialogue. Future multi-stakeholder processes will have to grapple with issues of organisational versus individual representation, if they follow the advisory forum model. There is a strong case for promoting more equitable women's and men's participation in such a design.

The Funding Challenge: Ensuring Adequacy and Independence

The WCD's fundraising approach raises three questions. First, from a pragmatic point of view, did the WCD raise adequate funds to sustain itself effectively? Second, did the process of fundraising support or undermine the commitment to inde-

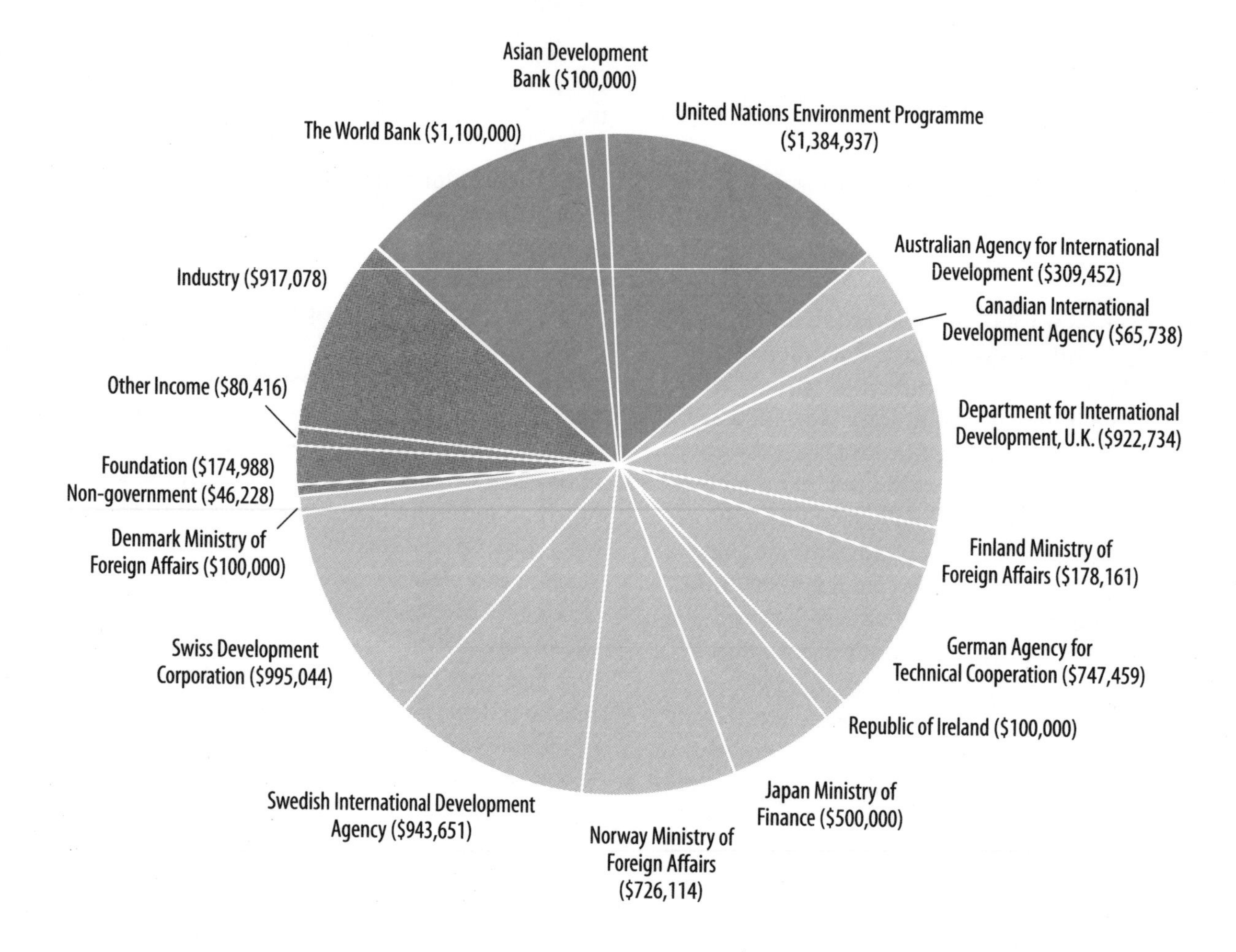

Figure 4.2 Income sources

Source: World Commission on Dams Project and Financial Report, May 1998 – April 2001.

pendence, broad participation, and transparency that characterised the WCD's work? Third, did the funding sources constrain or inhibit the working of the WCD?

From a pragmatic perspective, the WCD successfully met an ambitious budget, which allowed it to accomplish a significant amount of work over its two-year duration. This was a commendable accomplishment. At the same time, the WCD laboured under considerable uncertainty, particularly in its early months, over whether it would meet its fundraising targets. This uncertainty hampered its early work. Greater certainty of funds in the planning phase would undoubtedly have facilitated smoother implementation of the work plan.[40]

The fundraising strategy was designed to ensure the WCD's independence from possible sources of influence over its product and to mirror the multi-stakeholder nature of the Commission itself. Independence was sought by adopting a "no strings attached" fundraising policy. To ensure participation from a spectrum of actors, the WCD sought funds from the public sector, the private sector, and civil society institutions. The intent was to avoid undue reliance on any single funder and promote buy-in by all interest groups in the debate.[41] The total projected budget for the WCD was just under US$10 million.

One indication of the fundraising strategy's success was that the sources of funds and the means of raising funds did not provoke any debates during what was a highly scrutinised process. Multiple sources of funds allowed the WCD to claim that there was broad belief in the work of the Commission, as evidenced by the range of funders it attracted, including governments, industry, multilateral organisations, NGOs, and private philanthropic foundations. In examining the funding strategy in more depth, however, two caveats emerge.

First, although a wide range of funders from a variety of sectors did contribute, the bulk of the contributions came from the public sector. As Figure 4.2 shows, governments and multilateral agencies accounted for 88 percent of total financing for the WCD. Bilateral donor agencies from OECD countries, who funnelled their contributions through a trust fund set up by the World Bank, were a particularly important source of financing. This suggests that other stakeholder groups cannot or will not pick up the tab for commissions of this nature.

Second, the process of fundraising, at least in part, drove expansion of the Forum. Several Forum members, including a multilateral organisation, a government agency, and a private sector firm reported that they joined the Forum after being approached for a financial contribution to the WCD.[42] Did this compromise the independence of the Forum? The answer is likely negative, because there is no evidence that funding was a requirement for membership on the Forum for these groups. Indeed, in one case, the organisation, citing its internal constraints, refused to fund the WCD but was nonetheless invited to participate on the Forum. However, the fact that some Forum members did perceive a linkage between Forum membership and funds suggests that the Secretariat did have trouble maintaining absolute independence.[43] This process does illustrate the potential for fundraising imperatives to compromise the independence of a commission, or, almost as serious, to create perceptions of compromised independence, which could damage the integrity of the process.

Second, the WCD made some compromises on its "no strings attached" clause. Specifically, in some cases donors made requests regarding how their funds would be used, or provided in-kind information or expertise. Normally, this had no discernible impact on the functioning of the WCD. For instance, the U.K. Department for International Development (DFID) sought to support British input to the WCD. The WCD was able to meet this requirement through internal accounting that showed DFID funds were used to support British citizens on the Secretariat.[44] The Asian Development Bank provided funds earmarked for the WCD's consultations in Asia. Such geographically bounded provisions are not uncommon for bilateral and multilateral agencies.

Occasionally, funder preferences did influence the work programme in small ways. For example, the Inter-American Development Bank (IDB) chose to provide an in-kind contribution to the survey portion of the work programme from a Norwegian trust fund under its control. From the perspective of WCD staff, this led to considerable IDB control over the process, as the IDB, not the WCD, chose a Norwegian consulting firm to conduct the work and added questions to the survey. For its part, the IDB suggested that they were forced into more of a management role by the WCD than they desired.[45] In another case, a potential funder was rejected because the institution wanted to tie its funds. It subsequently dropped the restrictions and joined.[46]

The fundraising strategy was designed to ensure the WCD's independence from sources of influence.

In summary, the WCD faced a trade-off between independence and obtaining sufficient funds to promote inclusion in the process. The evidence does not suggest that these incidents substantially altered the trajectory or outcomes of the WCD, nor were they significant in undermining perceptions of the WCD's independence. However, they do suggest that the WCD faced continuous pressures in maintaining independence, and that the WCD had to make judgement calls as to what degree of interference was acceptable.

Conclusions

A commission built in part around the representation of stakeholders provides a promising alternative to a commission based on eminent persons. If key stakeholder groups perceive commissioners as legitimate representatives of their interests, then such a structure increases the likelihood that stakeholders will endorse the final outcome of the process. In order for this model to work, however, stakeholder groups must have a voice in the process of selecting Commissioners, and this voice should be recognised early in the process. This was the case in the WCD's formation, when a small but diverse group of stakeholders from the Gland meeting participated in Commissioner selection.

Strategies for building credible multi-stakeholder processes

- Base representation on broad constituencies and skills-based categories, rather than on eminence alone, to create the political space for a large range of stakeholders to get involved.
- Undertake an assessment to determine major categories of stakeholders who must be brought to the table.
- Engage the full range of stakeholders early in the process of selection to gauge the political acceptability of commission composition, particularly if the commission is based on the representation of interest groups.
- Ensure that the composition of the Secretariat embraces disciplinary breadth and is seen to reflect broader stakeholder interests.
- Ensure that a gender perspective is represented in all of a commission's bodies.
- Seek diverse funding based on untied funds.

A model designed around representation also brings with it complexities of implementation. Representation of large, unbounded groups of stakeholders with no formal structures of co-ordination or accountability cannot be based on formal representation. Instead, representation is based on the loose accountability of commissioners and on their reputation as individuals. This hybrid model requires that each stakeholder group is able to build consensus around legitimate representatives. Advocacy NGOs and social movements felt well represented on the WCD because their high level of organisation led to coherent demands for representation. Industry groups, by contrast, were poorly organised to participate at first, and did not feel well represented. Finally, the WCD experience shows that representation along the lines of interest groups has to be integrated with representation of experience, regional diversity, and gender diversity.

The secretariat that administers the day-to-day functioning of a commission has a key role in creating the political space for diverse participation. If it is perceived as broadly representative of stakeholder interests, it will have the trust of these groups. In the case of the WCD, the Secretariat mirrored the make-up of the professional development bureaucracy. The availability of qualified staff can pose a challenge to achieving sufficient diversity across interests, expertise, and gender and in regional balance. However, diversity in a secretariat, no less than in a commission, is worth striving for.

A fundraising strategy that draws upon a wide variety of contributors and that explicitly seeks independence from funders is an appropriate approach for a multi-stakeholder commission such as the WCD. The fact that the WCD's funding was not a contentious issue and did not cause any interest groups to leave the process suggests that securing funding from diverse sources was an important part of the platform for broad stakeholder engagement. Not only did this strategy ensure the direct buy-in of many actors, it also built the trust and co-operation of others (for instance, had the WCD been only corporate-funded, it would have eroded the trust of civil society). Diversifying sources is an appropriate strategy to minimise dependence and possible control imposed by a narrow donor base.

Endnotes

1. World Bank/IUCN, *Large Dams: Learning from the Past, Looking to the Future* (Gland: IUCN, 1997).
2. World Bank/IUCN, 1997, p. 11.
3. Interview with participant at Gland meeting, 1 February 2001.
4. World Bank/IUCN, 1997.
5. Interview with Forum member, June 2000.
6. Interview with industry representative on the Forum, 27 February 2001.
7. Patrick McCully, "How to Use a Trilateral Network: An Activist's Perspective on the World Commission on Dams." Paper presented at Agrarian Studies Program Colloquium, Yale University, 19 January 2001.
8. "NGO Letter on the Draft Final List" sent by e-mail to IUCN/World Bank Co-Chairs of IWG, Senior Advisor to IWG, and Chair, WCD, 28 October 1997.
9. "NGO Letter on the Draft Final List" sent by e-mail to IUCN/World Bank Co-Chairs of IWG, Senior Advisor to IWG and Chairperson, WCD, 28 October 1997.
10. Letter by Kader Asmal, John Briscoe, and George Greene to the Reference Group, 21 November 1997.
11. Interview with former IWG member, 16 November 2000.
12. Interview with former IWG member, 16 November 2000.
13. Interview with Commissioner, 27 February 2000.
14. Interview with Commissioner, 8 December 1999.
15. Interview with Commissioner, 27 February 2000.
16. Letter from John Briscoe and George Greene to the Reference Group, 23 December 1997. The Co-Chairs proposed the final date for the launch of the Commission to be late January 1998.
17. Letter from John Briscoe and George Greene to the Reference Group, 23 December 1997.
18. Based upon interviews with government and agency officials at the WCD consultation in Egypt, December 1999, and focus groups and interviews in Tanzania, Kenya, and Uganda in November 2000.
19. Interview with Commissioner, 8 December 1999.
20. Interview with Gland participant, April 2000.
21. Interview with industry representative on the Forum, 26 February 2001.
22. Interviews with industry representatives on the Forum, 17 November 2000 and 26 February 2001.
23. Wolfgang Pircher stated his reason for withdrawal as financial: Commissioners were not provided with consultancy fees.
24. However, this view was volunteered in an interview late in the process, after Dr. Veltrop had shown himself to be remarkably open to a wide range of views. Interview with Forum member, 17 February 2001.
25. Interview with official of India's Central Water Commission, speaking in his personal (non-official) capacity, 6 December 2000.
26. McCully, 2001.
27. Interview with Secretariat staff, 6 November 2000.
28. Interview with advisor to Kader Asmal, April 2000.
29. They argued that the short time frame within which the Commission was set up and the requirement to move to Cape Town limited the choice to development consultants whose experience drew largely from the international development bureaucracy. Interview with NGO representative on the Forum, 5 April 2000.
30. Interviews with Secretariat staff, 8 April 2000 and 26 February 2001.
31. Interview with Secretariat staff, 6 November 2000.
32. Interview with Secretariat staff, 6 November 2000.
33. World Bank/IUCN, 1997, p. 10.
34. Statement by Kader Asmal at WCD second Forum meeting, Cape Town, April 2000.
35. WCD website, www.dams.org/about/forum_list.htm (28 September 2001).
36. Felix Dodds, "Multi-Stakeholder Processes in the Context of Global Governance," in *Multi-Stakeholder Processes for Governance and Sustainability Beyond Deadlock and Conflict*. Minu Hemmati, ed. (London: Earthscan, 2001). Thanks to Dr. Hemmati for her insightful comments on this issue. Personal communication, 29 July 2001.
37. Personal communication with industry representatives on the WCD Forum, 26-27 February 2001.
38. Principle Three of *The Dublin Statement*, International Conference on Water and the Environment: Development Issues for the 21st Century, 26-31 January 1992, Dublin, Ireland. "Women play a central part in the provision, management and safeguarding of water: This pivotal role of women as providers and users of water and guardians of the living environment has seldom been reflected in institutional arrangements for the development and management of water resources. Acceptance and implementation of this principle requires positive policies to address women's specific needs and to equip and empower women to participate at all levels in water resources programmes, including decision-making and implementation, in ways defined by them."
39. Kalpana Sharma, "Lack of Rancour Marks WCD Proceedings," *The Hindu* (13 December 1998). Online at: www.dams.org/media/mediaitem.php?item=10 (28 September 2001).
40. Interview with Secretariat staff, 13 December 2000.
41. WCD website, www.dams.org/about/funding.htm (28 September 2001).
42. Interviews with Forum members, 1 December 2000, 27 February 2001, and 7 March 2000.
43. Interview with Forum member, 27 February 2001.
44. Interviews with Secretariat staff, 3 November 2000.
45. Interview with IDB official, 1 December 2000.
46. Personal communication with former Secretariat staff, August 2001.

Chapter 5

Implementing the Work Programme: The Commissioned Studies

As with other multi-stakeholder processes, the success of the World Commission on Dams rested on its legitimacy with the stakeholders whose actions brought it into being, and to whom it would turn over its findings. Indeed, the final product's authority depended upon a good process that enabled diverse stakeholders to contribute. The implicit benefits of diverse engagement were two-fold: first, such a process would be better informed by integrating diverse subjective viewpoints. Second, inclusion would build constituencies for implementation.

In order to "get the process right to ensure legitimacy," [1] the Commission committed to a set of guiding principles for its work programme. These included transparency, inclusiveness, independence, and accessibility. This chapter and the following chapter ask: How did the WCD put these principles of good governance into practice? What was the effect of these efforts on stakeholder buy-in to the Commission's work? We also consider the implications of this experience for the design of future processes.

An Inclusive Approach to Knowledge Gathering

Participants at the Gland workshop articulated what the Commission should do, and how to do it. They called for a Commission that would make decision-making on dams more transparent and accountable[2] and that would model these values in its own practice.[3] The challenge of operationalising this mandate rested with a group of very diverse Commissioners, brought together through an acrimonious process. Forging a work programme to the mutual satisfaction of such a group brought greater challenges than if the Commission had been less heterogeneous and less closely tied to the interests in the dams debate.

The Commissioners wisely decided on a flexible work programme that created the political space for diverse views on dam planning and practice to emerge. They used the creation of a global "knowledge base," a central part of their mandate, as an opportunity to engage most major interest groups. The next section of this chapter details aspects of the work programme's design that were intended to capture these diverse views and engage a broad cross section of stakeholders in the knowledge-gathering exercise.

Defining the specific tasks and scope of the knowledge base was a daunting task. Given the huge number of large dams—45,000 in all—the Commission realised it would not be possible to analyse a statistically representative sample. Rather, according to early work documents, the Commission would foster "a structured, transparent, and inclusive dialogue around key issues and major controversies that have proved to be so divisive."[4]

The dialogue would be accomplished through four activities, which became known as the "four pillars" of the Commission's knowledge base and are detailed in Box 5.1.[5] First, in-depth case studies would illuminate large dams' performance. Second, a survey of 150 large dams would capture trends in performance. Third, cross-cutting issues papers ("thematic reviews") would highlight best practices and recurring problems from around the world, as well as alternatives to large dams in providing water supply, energy, and flood control services.[6] Fourth, public consultations would be held in all major world regions to provide stakeholders with the opportunity to share their views directly, and the public would be able to make

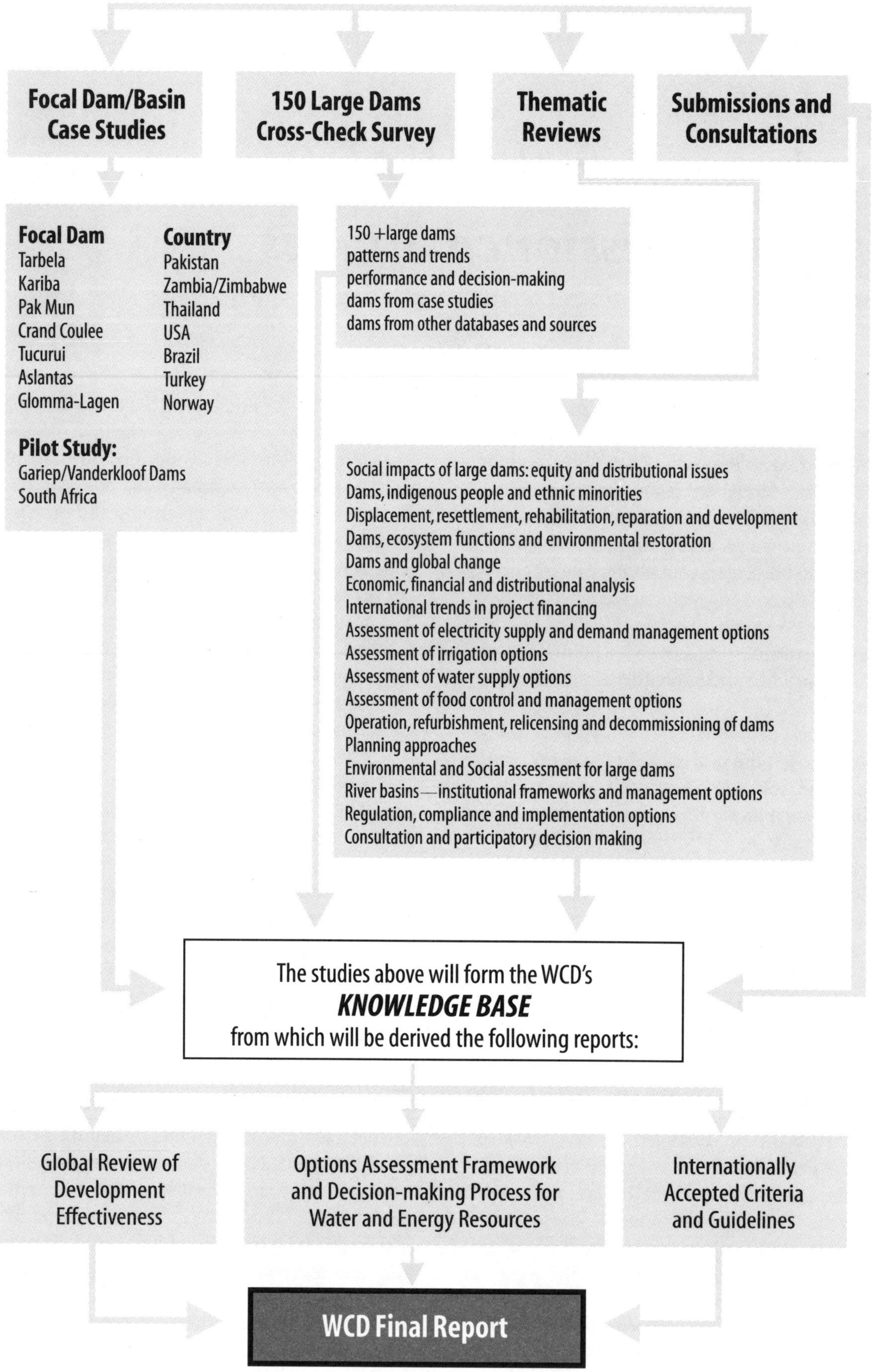

Focal Dam/Basin Case Studies

150 Large Dams Cross-Check Survey

Thematic Reviews

Submissions and Consultations

Focal Dam	**Country**
Tarbela	Pakistan
Kariba	Zambia/Zimbabwe
Pak Mun	Thailand
Crand Coulee	USA
Tucurui	Brazil
Aslantas	Turkey
Glomma-Lagen	Norway

Pilot Study:
Gariep/Vanderkloof Dams
South Africa

150 +large dams
patterns and trends
performance and decision-making
dams from case studies
dams from other databases and sources

Social impacts of large dams: equity and distributional issues
Dams, indigenous people and ethnic minorities
Displacement, resettlement, rehabilitation, reparation and development
Dams, ecosystem functions and environmental restoration
Dams and global change
Economic, financial and distributional analysis
International trends in project financing
Assessment of electricity supply and demand management options
Assessment of irrigation options
Assessment of water supply options
Assessment of food control and management options
Operation, refurbishment, relicensing and decommissioning of dams
Planning approaches
Environmental and Social assessment for large dams
River basins—institutional frameworks and management options
Regulation, compliance and implementation options
Consultation and participatory decision making

The studies above will form the WCD's
KNOWLEDGE BASE
from which will be derived the following reports:

Global Review of Development Effectiveness

Options Assessment Framework and Decision-making Process for Water and Energy Resources

Internationally Accepted Criteria and Guidelines

WCD Final Report

Box 5.1 **The four pillars of the knowledge base**

Source: WCD website, www.dams.org/about/workprog.htm (28 September 2001).

general submissions to the Secretariat, by mail, in person, or through the Commission's website. This framework emerged out of considerable discussion and after several alternative frameworks were considered.

The flexible work programme created political space for diverse stakeholder views.

The Commissioners did not review past dams experience with a specific yardstick for "development effectiveness" (indeed, it would likely have been impossible for the diverse Commissioners to agree on such a yardstick at the start). Instead, they encouraged stakeholders to air their own views on what constitutes development effectiveness, and they promised to weigh convergent and divergent views.[7] Kader Asmal's injunction for people to have their "day in court" in the context of one component of the work programme, the regional consultations, was therefore a suitable allegory for the WCD's work as a whole. Normative judgements about the justice of dam building and distribution of benefits were as welcome as technical cost-benefit calculations when it came to stakeholder consultations and general submissions.

Significantly, this approach gave many groups the hope that their views might prevail. If, by contrast, the Commission had tried to come up with a single yardstick for assessing dams, they might have alienated certain interest groups from the start. The Commission side-stepped the flaw in the World Bank Operations Evaluation Department (OED) study that had ignited NGO criticism. The OED Review had rated World Bank-funded dams on a scale of unacceptable to acceptable but had run into trouble because NGOs criticised the World Bank's evidence and OED's notion of "acceptable."[8]

"In many other areas, people feel excluded," said one Commissioner from government. "Here we have a unique opportunity, people feel included."[9] A community-based Commissioner rallied an NGO meeting with a similar statement: "I think we can reassure the communities that the WCD is demanded by the people themselves. You can tell them it's up to us, the civil society, as to how we use this process effectively."[10]

In spite of these benefits, it is important to note a discrepancy among stakeholders of what the knowledge base was meant to achieve. According to the Commission and Secretariat, the knowledge base was intended to both contribute to the Commission's learning from the past and to highlight current and future good practice in dam building, operations, and decommissioning. But government and industry actors, on the one hand, and NGO and movement groups, on the other, had fundamental differences about the appropriate orientation of the workplan. Government and industry actors thought the exercise should focus as much as possible on good practice in recent dams history. They were looking for changes that could be adopted relatively easily within given development frameworks. Non-governmental actors were looking for full documentation of bad dams practice that would support their campaigns for compensation for displaced people and their desire for large dams technology to be condemned in the future. This ongoing contestation would colour the next two years' work.[11]

The regional consultations gave people their "day in court."

The knowledge generation process for the WCD could have taken different paths. Early on, Commissioners discussed whether they could individually supervise independent reviews of dams experience in their regions[12] with light co-ordination by the Chairperson and Secretary-General, but the Commissioners rejected this model on the basis that their credibility rested on a uniformity of approach across case studies. This could only be achieved by having a substantial body of senior advisors in Cape Town to co-ordinate the studies.[13] Considering that individual Commissioners tended to raise suspicion when they appeared alone in the WCD context, it is also possible that such a decentralised model would have alienated key interest groups in the regions. Although rejection of this model implied hiring a larger Secretariat, it probably increased the inclusiveness of the process.

Some development practitioners and Secretariat members favoured a second model that involved undertaking a full-blown assessment of the development effectiveness of energy alternatives to hydropower.[14] This approach would have placed the future of dams within the framework of global energy forecasts (as well as freshwater needs), and would have illuminated the comparative advantages and disadvantages of dams over other means of obtaining these services. Many industry and governmental participants remain displeased, to this day, that such a comprehensive options assessment was not undertaken. The Commission rejected this model on the basis that global forecasting was a niche occupied by existing think tanks.[15] Moreover, a technocratic approach of this nature would not have supported the commitment to inclusion and participation demanded by the Reference Group at Gland.

Participation in the Framing Process

The earliest expressions of the Commission's mandate called for a body that would undertake its work in a participatory fashion.[16] Yet, the convenors of any multi-stakeholder process must ask: When should we invite stakeholders to participate? Should the process be open to public comment at every step of the way, or does this make the transaction costs too high? This question applied at every stage in the WCD process: from the call for a World Commission on Dams, through the drafting of the final report. As shown in Chapter 3, an open dialogue among core stakeholders on the composition of the Commission proved essential to obtaining their groups' support for the two-year process. In the following process to shape and define the work programme, the newly formed Commission erred toward providing frequent opportunities for public comment.

The catch with the WCD's initial efforts to invite participation is that they were not sufficiently publicised to garner significant input. Because of funding constraints and a small staff in the early days, the WCD's invitations for public comment on the emerging work programme were largely disseminated on its website in 1998. However, this virtual forum was barely used and was consequently removed from the site.[17] Instead, comment came predominantly from networks of NGOs and professional associations who learned of the WCD from colleagues on the Commission and Secretariat and used these contacts as pressure points.[18]

It was not until the first meeting of the WCD's advisory Forum in April 1999 that a broader range of stakeholder inputs—through the Forum members and their networks—developed. During this meeting, the diverse members of the Forum became fully informed about the scope and elements of the work programme. By then, the window of opportunity was closing for comment on the framing of Commission studies. Of this consultation period, one senior Secretariat member noted, "One lesson is, don't assume that if you don't have comments on the work programme that they're happy with it!"[19]

One Forum member from a development bank noted that the process of consolidating the work programme had happened too fast for his colleagues to absorb and respond to the information. "[The WCD] needed time for proper outreach on the methodology, for country studies and finding support from all corners," he said.[20]

In fairness to the WCD, its budget was extremely tight for the first two-thirds of its history, with scarce funds for elaborate outreach. The Secretariat's outreach efforts were challenged by the amount of time that the Commissioners took to settle on the precise elements of the workplan (it took until December 1998 to establish the main cases and questions). Outreach was also slowed by fundraising. Difficulties in fundraising distracted Secretariat staff from their other work and delayed portions of the work programme.[21]

The WCD's experience highlights the importance of publicising the nature and aims of a commission at the outset, extending beyond electronic means as much as possible. The WCD made reasonable efforts to reach out in person. Future commissions should try to do even more to ensure they are reaching those without Internet access, and to ensure they win relevant stakeholders' attention early.

The Commissioned Studies: Case Studies

The Choice of Case Studies: A Political Balancing Act

The WCD's case studies were a key component in the Commission's Global Review of Large Dams. They were intended to give "the first integrated look at dams from the perspective of all interest groups, be it from the point of view of government agencies, local economists, the riparian habitat, or impacts on the diets of indigenous peoples."[22]

The main question facing the Commission in its choice of studies was: How would it access an adequate breadth of experience about large dams and their impacts? It was not simply a question of how to assemble a representative picture of the experience with large dams, but also of how to ensure that the Commission appeared neutral to outside audiences. The Commission did not wish to alienate any major groups with its choice of case studies.

Dam proponents feared the Commission would choose only "failed" dams as a way of tarnishing the whole industry.[23] Dam opponents wanted to make sure that the WCD recognised some of the grossest human rights abuses, dam-related corruption, and questions of ecological and economic viability around which they had mobilised in the first place. Would the Commission demonstrate its neutrality by choosing a balanced set of case studies of "good" and "bad" dams?

The WCD appeared to satisfy most major stakeholder groups who were then monitoring the process—the advocacy NGOs and dams interest groups—by selecting a set of case studies that encompassed diversity in geography, function, age, size, and catchment area.[24] The case studies are listed in Box 5.2.

Dam proponents feared the Commission would look only at failed dams.

Box 5.2

The WCD case studies[a]

Focal Dams and River Basins:

Brazil Tucurui Dam and Amazon/Tocantins River
Norway Glomma and Lågen River Basin
Pakistan Tarbela Dam and Indus River Basin
Thailand Pak Mun Dam and Mekong/Mun River Basins
Turkey Aslantas Dam and Ceyhan River Basin
United States Grand Coulee Dam and Columbia River Basin
Zambia and Zimbabwe Kariba Dam and Zambezi River Basin

Country Reviews:

China
India
Russia

Pilot Study:

South Africa Gariep and Van der Kloof Dams and Orange River Basin

[a] The Commission intended for the case study dams to be set in a basin-wide context, as explained in the WCD's Work Programme of February 1999, so that they might illustrate the cumulative effects of a cascade of dams or the effects of dams far downstream beyond the project site. This proved difficult to accomplish in practice, and the study of the Glomma and Lågen Basins in Norway was the only study that demonstrated cumulative impacts and decision-making across an entire river basin. (Interviews with senior Secretariat staff, 6 November 2000 and 28 February 2001.)

Source: World Commission on Dams, Dams and Development: A New Framework for Decision-Making (London: Earthscan, 2000), p. 31.

An additional criterion for case study selection was diversity in political regimes, so that the Commission could study the differences in decision-making around large dams. *(See Box 5.3.)* However, in the pursuit of such political diversity, the Commission also wanted to avoid regimes so oppressive that it would be unable to consult with communities or raise transparency and accountability issues in the course of its assessment.[25] The WCD case studies did demonstrate political diversity, but restrictions on civil society participation in three of the countries (Turkey, Pakistan, and China) limited the vitality of discussions later on. The Commission faced a trade-off between analysing the diverse political conditions under which dams are built and its desire for a thorough study in each case.

The only major upset about the choice of case studies occurred when ICOLD, the main dam industry association, learned from an internal WCD document that Turkey's Ataturk Dam was on the shortlist.[26] Ataturk was a primary example of a "problematic" dam that the association feared the WCD would use to cast doubt on the profession as a whole. As some ICOLD members were already suspicious of the Commission, choosing the Ataturk Dam may have caused a serious breach with this group. The Secretariat hurriedly removed the dam from consideration.

Community-based organisations put pressure on the Commission to intervene in the decision-

Box 5.3

Guiding questions for the case studies

1. What were the projected versus actual benefits, costs, and impacts of the dam?
2. What were the unexpected benefits, costs, and impacts?
3. What was the distribution of costs and benefits - who gained and who lost?
4. How were decisions made?
5. Did the project comply with the criteria and guidelines of the day?
6. What were the lessons learned?

Source: World Commission on Dams, Dams and Development: A New Framework for Decision-Making (London: Earthscan, 2000), p. 30.

making related to controversial dams under planning or construction. This tension first emerged during case study selection but resurfaced many times during the process.

The Controversy over Assessing Current Dam Projects

Two community-based organisations pleaded with the WCD to look at their cases, in the hope that it would have a positive influence on their campaigns. The WCD declined to get involved because the projects were ongoing and the Commission judged their intervention to be politically infeasible. This wise decision enhanced the Commission's wider legitimacy.

The Brazilian Movement of Dam-affected People (MAB) encouraged the Commission to choose a particular dam where intensive civil society engagement had changed the features of the project.[27] They wanted to demonstrate what could be accomplished when decision-making processes were democratised. Meanwhile, the Cree Nation, a native people in Canada, asked the Commission to consider the Churchill Falls Dam in Newfoundland, which was being planned in the face of native people's dissent.

Many Commissioners feared it would be difficult to appear balanced and that they would risk alienating stakeholder groups. "The whole political space of the Commission would have been destroyed" if stakeholders had understood it to be adjudicating on current dam controversies, noted one senior advisor.[28] "One of the spaces the Commission has is to look at the range of experience of dams worldwide" without judging specific ones. Furthermore, most Commissioners believed that for the purposes of assessing development effectiveness in the case studies, it was necessary to study completed dam projects from which the benefits and costs already flowed, with all their foreseen and unforeseen impacts.[29]

In the case of the Brazilian social movement's and the Cree Nation's interventions, failure of the WCD to pick their choice of case study did not make these actors leave the process. However, the overall issue of whether the WCD would study current dam projects in depth continued to irk community-based and NGO stakeholders. Their dissatisfaction was understandable. Several groups with representatives on the Commission or WCD Forum were active in current struggles to block large dams or seek reparations from past projects. These groups justified their involvement in the WCD process with the hope that their commitments in scarce human and financial resources would bring progress in their specific campaigns.[30]

To satisfy these groups, the Commission had to seek less confrontational ways of allowing them to express concerns about current dam projects. For instance, stakeholders could make submissions to the Secretariat about current dam concerns and, in most cases, had the opportunity to present current issues at regional consultations (although this process was not entirely "free," as described in Chapter 6).

Stakeholders on the pro-dams side complained that the dams studied by the Commission were too old and did not adequately reflect the advances in environmental mitigation technology and compensation practices made in recent years. However, most of these complaints did not surface until after the WCD report was released. Ironically, NGOs' and peoples' movements also wanted the WCD to look at dams under construction in order to demonstrate what they perceived as the inadequacy of current decision-making processes and mitigation measures.

Later events supported the WCD's decision to distance itself from current dam projects. As documented in Chapter 6 *(see Box 6.3)*, the Indian government perceived the WCD to be meddling in the Narmada Valley Dams dispute when the Commission planned a field trip to the Valley. This conflict damaged relations between the WCD and the Government of India almost irreparably. The Chinese government was initially involved in the WCD but later withdrew from the process. One of the reasons for withdrawal, according to the WCD, was that officials of China's Ministry of Water Resources mistakenly understood the WCD to be passing judgement on the Three Gorges Project.[31]

Future commissions and multi-stakeholder processes are likely to encounter similar tensions around the discussion of controversial current projects. They would do well to learn from the WCD's wisdom and refrain from intervening, or being seen to arbitrate in, specific disputes.

The Country Studies: A Second Best Option

The Commission sought to be as comprehensive in its knowledge base as resources would allow. In some cases, lack of government co-operation limited available data. According to the Commission's final report, the WCD sponsored country-level studies of large dams in India and China because these governments would not agree to in-depth studies of individual dams and river basins.[32] This alternative approach was borne of the political tensions described above. In the case of Russia, another major dam-builder, the Commission could not raise the funds for a case study and settled for a more modest briefing paper.

Almost nobody was satisfied with the results of the China and India country studies. The Indian government's Central Water Committee was extremely critical of the study, because it felt its officials were inadequately consulted during the process. Civil society groups criticised the study for its lack of thoroughness and its slim treatment of options to large dams. Those in the international community who were knowledgeable about China's society and environment found little of use in the China study, which failed to address political economy issues in a significant way.[33]

The Commission's difficulty in maintaining the trust of the Indian and Chinese governments, a story in which the country studies play just one part, was to vex the Commission throughout its history. China's and India's distancing from the process held implications for the WCD's inclusiveness and demonstrated the hard reality of the political trade-offs the process faced. We revisit these questions in Chapters 8 and 9, where we explore broad stakeholder reactions to the WCD's final report, and the relation between the process and stakeholder willingness to promote and adopt the recommendations.

Difficulty in engaging the Indian and Chinese governments would vex the Commission throughout its history.

The Role of Case Study Teams in Creating an Inclusive Knowledge Base

The Commission instructed its case study consultants to seek quantitative and qualitative information to "assess and illustrate stakeholders' views."[34] To accomplish this, consultants' ability to seek out diverse views and garner the respect of different stakeholders was critical.

According to the WCD's final report, the Commission decided to employ "national teams of experts rather than using international consultants...while creating greater challenges in terms of independence and neutrality it provided the Commission with a deeper insight into the political, historical and cultural contexts for water and energy resources management."[35] Forum members welcomed the choice of national teams, for many members (especially NGOs) suspected that international consultants would treat local problems superficially. However, as the Commission acknowledged, there were significant practical challenges to assembling study teams that were politically acceptable to most stakeholders.[36]

The Secretariat sought study teams with multidisciplinary expertise and from a range of institutions in the relevant country to provide a multistakeholder profile. Most of all, the Secretariat sought consultants who were open-minded and had experience working in different sectors, such as government, NGO, and business.[37] This strategy had mixed success in soliciting information from the broader community of stakeholders. But the strategy was successful enough to suggest that the model is worth trying in future commissions.

Box 5.4

The World Commission on Dams and China

Initial efforts by the WCD and its convening institutions to court the Chinese government reflected China's dam-building status. The only ministry-level representative at the Gland meeting came from China. A representative of the country's Ministry of Water Resources was offered a place on the Commission, and Shen Guoyi accepted. The government agreed to sponsor the WCD.[a] Initial discussions took place between the Commission and the Chinese government about the possibility of undertaking a case study of the Danjiangkou Dam.[b]

It is difficult to find public sources of information about the Chinese government's view of the WCD. However, early articles by Chinese officials indicate that they welcomed these opportunities to summarise past experiences and study the dam-building practices of other countries. A working group of Chinese experts was established to provide opinion on the WCD's studies, write a report on China's position on dams and sustainable development, and prepare materials for members of the Chinese National Committee on Large Dams (CHINCOLD), who intended to participate in WCD activities.[c]

In time, China's engagement with the WCD tapered off. As the WCD's requests for technical data became more detailed, the government became less responsive. Almost one year into the process, China declined permission for the WCD to undertake a full case study of the Danjiangkou Dam.[d] Around the same time, a restructuring occurred in the Ministry of Water Resources that coincided with decreasing political support for WCD activities, including a suspicion that the WCD was against dams and that its discourse was incompatible with Chinese interests.[e] Ms. Shen resigned from the Commission, citing health reasons. The Commission downgraded its assessment of China's dams experience from a case study to a country study, and then to an external review undertaken by foreign consultants.

In the final analysis, of more than 180 consultants hired by the WCD in its work programme, only 3 were Chinese. There was no Chinese national on the Commission, Secretariat, or Forum following Ms. Shen's resignation. The government refused to release basic technical data on its 22,000 large dams, which represent almost half of the global population.[f] China's involvement in the WCD had effectively ended.

According to a senior Secretariat staff member, the WCD continued to send materials to their "many contacts in the [Chinese] system."[g] However, once the Commission lacked official sanction, its outreach to Chinese institutions was cautious and limited. China's representation at the WCD's East Asia regional hearing in Vietnam was extremely modest. In China, media reports came to be dominated by resounding approval of the Three Gorges Project provided by ICOLD engineers during their September 2000 Congress in Beijing. WCD Commissioner Judy Henderson's presentation to the ICOLD Congress at the time received scant attention.

The Chinese government's withdrawal from the process, although difficult to corroborate from direct sources, may be analysed within the complex framework of China's current political situation. There is growing awareness in China that more is at stake in the Three Gorges Project than isolated technical and social issues. Consequently, internal tension over the feasibility and possible impacts of Three Gorges has influenced the Chinese government's approach toward dams issues, both domestically and internationally.[h]

---

[a] Meaning that the Chinese government provided an in-kind donation of officials' time to the data gathering process. E-mail correspondence from former WCD Secretariat member, "China and the WCD," 17 April 2001.

[b] Interview with Commissioner, 19 March 2001; WCD website, www.dams.org/about/wp_ov_anx1.htm (28 September 2001).

[c] Zhu Dangsheng, "Daba yu huanjing wenti" (Dams and Environmental Issues), Ministry of Water Resources document. Online at: www.dppr.com/txt/a02.htm (28 September 2001).

[d] E-mail correspondence from former WCD Secretariat staff, "China and the WCD," 17 April 2001.

[e] E-mail correspondence from former WCD Secretariat staff, "China and the WCD," 17 April 2001. Also, interview with senior Secretariat staff, 6 November 2000; World Bank, internal document.

[f] Interview with senior Secretariat staff, 6 November 2000.

[g] Interview with senior Secretariat staff, 6 November 2000.

[h] See, for example, "Three Gorges Dam Project," Trade and Environment Database (TED) Case Study no. 264, American University. Online at www.american.edu/ted/THREEDAM.htm (28 September 2000). "Dam politics: How Three Gorges plays in Beijing," Asia Times Online, 5 May 2000. Online at: www.atimes.com/china/BE05Ad01.html (28 September 2001).

Source: Literature review drawn from Fredrich Kahrl, "Under the Shadow of the Three Gorges Dam: The World Commission on Dams and China." Background paper prepared for the WCD Assessment, January 2001.

An American team was the most successful study team in eliciting diverse views from the local and national experience and producing a report with substantially new information. The team leaders were civil engineers from two well known universities who were considered politically neutral by stakeholders at the study site. They sought a range of complementary expertise in economic, ecological, and social issues among colleagues in other university departments. Under criticism from at least one Commissioner for not having someone with practical experience on the team, they later added a consultant from Harza Engineering Corporation. The team also took care to survey local communities, including Native American peoples, for data on the demonstrated costs and benefits of the Grand Coulee Dam. Their report represented convergent and divergent areas of stakeholder opinion. A wide range of concerned local parties praised the report.[38]

An ambitious effort to involve a diverse collection of scholars and practitioners in co-authoring the case study in Thailand, including a radical academic, consulting engineers, and government fisheries staff, was difficult to manage in practical terms, perhaps because of the sheer range of perspectives.[39] The team nonetheless succeeded in producing a path-breaking report that documented how the Pak Mun Dam had affected fishing communities' livelihoods on an unforeseen scale and had failed to pay for itself. The World Bank and the Electricity Generating Authority of Thailand (EGAT)—the Pak Mun Dam's backers—rejected the findings on the basis that the authors did not use an appropriate model to forecast the dam's rate of return. The Bank and EGAT never shared the relevant models with the WCD or its consultants as proof.[40]

In the Pakistan study, the study team came predominantly from one development consultancy, but local stakeholders disputed the team's political neutrality. Team members from Asianics Agro-Dev International had a variety of expertise, including irrigation, agricultural economics, sociology, and environmental science. One of the chief criteria for choosing the firm was its apparent independence from government.[41] However, civil society groups considered the team to be strongly biased toward the dams establishment.[42] To redress the balance, the WCD invited one of the main critics, a non-governmental advocacy group called the Pakistan Network on Dams, Rivers and People (PNDRP), to contribute staff toward the effort. They declined to join the analysis, but were instrumental in mobilising affected people and others to take part in the review meeting, which had a significant effect on the content of the final draft.

Based on this diverse implementation experience, was there a winning combination of characteristics in a case study team to ensure broad stakeholder acceptance and, hence, credibility for the Commission? The most accepted combination appears to have been an academic/research team that combined interdisciplinary expertise with overall political neutrality. Having some practical experience on the team in managing large dams or their impacts was necessary to gain the trust of practitioners, and having experience with, and appreciation for, project-affected peoples was necessary to gain the confidence of NGOs and social movements.

Review Meetings as a Means to Broaden Participation

A principal design feature for inclusiveness and transparency in the case studies were the multi-stakeholder review meetings. The first stakeholder meeting was intended to gather feedback on the terms of reference prepared by the Secretariat. The second multi-stakeholder meeting was intended to solicit comments on the consultants' draft of the case study and gather participants' oral and written views on the development effectiveness of the dam.

According to a senior Secretariat staff member, the case study meetings were the Commission's chance to get close to dam-affected and other local people. For example, the WCD arranged for Tonga chiefs from Zambia to make a long journey to attend a stakeholder meeting for the Kariba Dam study,[43] and it mobilised tribespeople on the Zimbabwean side and various NGO supporters.[44] The study teams who managed to mobilise credible community participation in these meetings earned the approval of international NGOs and agency personnel who were monitoring the process—and provided good publicity for the Commission. Many Forum members appreciated the WCD's efforts to go beyond the relatively elitist consultations of most regional and international policy processes. *(See, for comparison, Chapter 6, Box 6.1.)*

The inclusiveness of the dialogue at these review meetings depended on several factors outside the Commission's immediate control. It depended on the co-operation of the government and the overall enabling environment for civil society mobilisation and expression. Where the political conditions for NGO mobilisation were constrained, the Commission was unable to elicit a broad range of stakeholder viewpoints. For instance, the case study of the Aslantas Dam and Ceyhan River Basin in Turkey did not probe social and environmental issues in depth and failed to investigate the development benefits or losses to the displaced population. The country meeting to discuss the draft paper was dominated by the State Hydrological Works Department (DSI), whose officials refused to discuss resettlement and ethnic minority issues. Environmental and advocacy NGOs were entirely absent from the meeting, perhaps because of the weakness of the NGO sector in Turkey.[45]

Multi-stakeholder review meetings allowed communities to provide feedback on case studies.

Meanwhile, Pakistan's change from a nominal democracy to a military dictatorship while the WCD was undertaking the Tarbela Dam case study worried Commissioners. "In Pakistan, because of the military regime and the removal of the carpet of democracy and the denial of a healthy social and political process, we were very concerned about the lack of democratic participation in the Karachi [first stakeholder] WCD meeting," said one Commissioner. "Despite that, we decided to go ahead. While some serious problems prevailed, we were able to generate some open debate on a highly secretive and undemocratic process. Even the government and the army conceded that there was need for debate on large water management projects."[46] Similar concerns about participation under authoritarian regimes pervaded some regional hearings as well. These are described in Chapter 6.

The Commissioned Studies: Cross-check Survey of 150 Large Dams[47]

Transparency and Inclusiveness in the Cross-check Survey

The Commission complemented the in-depth case studies with a cross-check survey of the technical, social, environmental, and decision-making characteristics of 150 large dams around the world. The cross-check survey, as it was known, presented an opportunity to "expand on the case study dams and at the same time, to make [the data set] regionally reflective."[48] Although it could not claim to be statistically representative, the sample would "seek to generate broader patterns and trends" than was otherwise possible with the case studies.[49] Survey dams were chosen from the Commission's focal river basins as well as from existing databases (such as the World Bank's). Other dams were added to increase the sample's diversity. Through much of the two-year process, the survey was something of a poor cousin to other elements of the work programme: it had a low public profile and was slow to show results given the tremendous logistical challenges of gathering completed questionnaires.

The survey was an important source of independent data for the WCD's Global Review of Large Dams, as is evident in the Commission's final report. The primary method was a survey about the dam's technical, economic, social, environmental, and decision-making history distributed to dam operators, consultants, and research institutes. The survey did not have multi-stakeholder input built into its design, as with the case studies (resources did not stretch that far), but in the later stages of the work programme, the Secretariat conducted a limited review process to validate data and solicit wider input. The review process pleased NGOs that were monitoring the process, as it gave them a role. The Secretariat chose participants, mostly from local NGOs, to review 17 randomly chosen and 18 controversial dam projects in the total sample.[50] An activist NGO in Cape Town even became involved in contacting local NGOs in various countries and drumming up support for alternative contributions.[51]

Comprehensive responses to the survey were only achieved through an immense logistical effort by the Secretariat—a challenge that holds implications for future processes. Staff created software to help respondents complete the form. However, many of the respondents had difficulty downloading attachments or were simply too worried about computer viruses to use the programme. In the end, the Secretariat faxed and phoned extensively to gather the data they sought. Language problems also posed a challenge, as few respondents had English as a first language and translation was limited to a Spanish version of the survey. Dam operators' sheer lack of information about the selected dams also challenged the breadth and excellence of the sample. As the Secretariat noted, they could have chosen the "largest most controversial dams, and got a lot of data. (But) the fact is that most large dams are less than 30 meters in height. The interesting thing...was to see the impact of all these smaller dams."[52] Some governments did not even know the co-ordinates of the large dams selected for the Commission's survey.[53]

A cross-check survey of large dams sought to reveal broad patterns and trends.

This large push to gather responses was worth the effort because of the data on trends it generated for the final report. Certainly, no smaller global survey would have passed the credibility test with stakeholders. Reactions to the WCD's final report, given in Chapter 8, include criticisms that the survey was too small. Stakeholders from government, in particular, hoped such a survey could encompass their country's best practices. Such arguments had less to do with the success of the WCD's stakeholder engagement—the focus of this assessment—than with the technical merits of the WCD's methodology, which were in this case constrained by time and funding. The Commission surveyed as many large dams as it had time and money for, given the Gland mandate for a time-limited process, the Commissioners' need to bring closure to the knowledge-gathering exercise, and the fundraising challenges. Future commissions might face similar trade-offs between comprehensiveness and time and funding pressures, depending on whether the issue at hand requires data to be gathered for the first time from diverse original sources. It is not clear that critical stakeholders would have been any happier with the results if the WCD had been more comprehensive in its cross-check survey.

Conflicts over Knowledge

Even if data management had not been an issue, the WCD would have been challenged to nurture relations with the development agencies, professional associations, and technical and research institutes that hold the fragmented and (for the most part) poorly organised data concerning the performance of large dams. Negotiations for data can take time under the most open political and institutional regimes. In the WCD's case, staff and consultants also had to overcome potential contributors' scepticism or disinterest in the Commission's work and persuade them that the WCD was a worthwhile enterprise for the future of the industry.

Among the most important repositories of information about dams and dam-related development were the professional associations: the International Commission on Large Dams (ICOLD) and the International Committee for Irrigation and Drainage (ICID). Although these organisations eventually shared their data with the WCD, members remained sceptical that the WCD had any value to add beyond the professional datasets and standards their organisations had already developed. The contention over validity of datasets and analytic methods was captured by a senior ICID official's remark toward the end of the WCD's knowledge generation process: "We have the best databases on irrigation in the world. The WCD is not helpful, we are only in it [the Forum] for damage control."[54] Dam proponents unhappy with the WCD's findings would later use the issue of data validity and representativeness as a reason to dismiss the final report.

Governments were wary about the WCD's access to and use of dam-related data.

Governments also proved to be wary about the WCD's access to and use of material. Although the WCD emphasised that it was seeking to establish trends for the entire 150-dam sample, governments and utilities opposed public disclosure of raw data on individual dams. "Governments and utilities were worried that opponents would use [the data] against them," said the Secretariat member in charge.[55] In all, the WCD received permission to make public the individual raw data on only 50 dams out of the eventual 125 in the full survey. These constraints, while representative of the political realities of data disclosure in many countries, nonetheless undermined the Secretariat's commitment to greater transparency.[56] The sheer difficulty of extracting geographically diverse data on large dams suggests that future commissions and multi-stakeholder processes must plan for a time-consuming effort if they are to consolidate even existing data on a sector.

The Commissioned Studies: Thematic Reviews

The Commission's Terms of Reference, as articulated in Gland, said the Commission must respond to a changing global context where there are multiple criteria for decision-making. In addition, the WCD was born of the growing appreciation that past decision-making has emphasised the benefits of dams and ignored or underestimated the costs of dam building. The WCD, then, was to focus on "those key issues around which there is greatest disagreement."[57]

The thematic reviews focussed on issues around which there was greatest disagreement.

To look at issues of disagreement while satisfying all stakeholders of their impartiality was a considerable challenge. The Commission, based on drafts prepared by the Secretariat, agreed to pursue 17 thematic reviews on a range of controversial issues around dams. These reviews were grouped into five clusters of issues: social, environmental, economic and financial, and institutional and governance issues, and options for water and energy generation. In the Secretariat's view, technical organisations, such as ICOLD, had already covered more technical issues related to dams, and focussing on areas of controversy would allow the WCD to pursue its comparative advantage. However, this approach quickly rang alarm bells with some stakeholders. Industry groups active in the WCD process felt that the thematic reviews' focus on controversial issues would be unlikely to capture adequately the full benefits of dams. From their perspective, case studies of individual dams would better allow for a balanced assessment of costs and benefits.[58] For their part, civil society groups were convinced that a truly independent and objective review of experience with large dams—whether on a sectoral or case study basis—would vindicate their views. Hence, the WCD's credibility as a fair and neutral body was at stake in how the thematic reviews were carried out. What quickly became apparent, however, was that a research process of this nature could not be entirely free of negotiation with stakeholder groups. The Commission was tested on how it managed the stakeholder debates that inevitably arose over framing and content of the thematic papers.

Stakeholders' Jostling for Position

The thematic review process was designed to incorporate a substantial measure of transparency and openness. The Secretariat circulated terms of reference to reviewers for comment and placed them on the WCD's website. Drafts of the thematic papers were circulated to Commissioners and external reviewers, including Forum members. The shifting scope and definitions of thematic review studies somewhat hampered these efforts at transparency. For example, the review of Regulation, Compliance, and Implementation was narrowed from a comprehensive review of existing criteria, guidelines, standards, and legal, policy, and institutional frameworks for dams, to a more limited subset of these issues based on discussions within the Secretariat.[59] Reviewers lacked clear and timely signals about the status of the studies because of uncertain budget allocation for thematic review studies.

In addition, achieving agreement on the basic research question for several thematics proved to be a politically charged task. For example, the Economic, Financial, and Distributional Analysis thematic was hamstrung by a debate over the relative merits of a focus on theory, practice, or a review of past performance. *(See Box 5.5.)* Stakeholder groups clashed over the basic methodology of comparing ecosystem impacts of dams against a base scenario of no dam on the same site in the Ecosystems thematic review, with industry representatives rejecting this notion.[60] In one of the most intensive thematic review processes, debate over the appropriate scope and framing question for the Social Impact of Large Dams review continued well into the drafting stage. *(See Box 5.6.)*

Resolution of suspicions of this sort required implicit negotiation, which operated through submissions and comments to the Secretariat on the draft review papers. In the case of the social impact paper *(see Box 5.6)*, a special workshop provided the setting for explicit negotiation. To the Secretariat's credit, there was considerable scope for commenting on the papers overall. A core set of stakeholders from industry and civil society took advantage of these opportunities and were extremely active in scrutinising and commenting on draft papers. The engagement of these core actors enhanced the Commission's continued legitimacy with broader networks.

Some stakeholders from industry and civil society were extremely active in scrutinising papers.

Not all stakeholders were equally well equipped to take advantage of this mechanism for feedback. The Secretariat perceived civil society groups and academics as being better able to network and react quickly than were stakeholders used to functioning in a more bureaucratic manner, such as dam-builders' associations.[61] Governments participated in the review process via individuals from various government departments. They did not establish a collective, organised effort to influence the framing of issues as did civil society and industry groups. As a result, government representatives had a less sustained and influential voice in shaping the thematic papers.

Commissioners reviewed the terms of reference for the studies, but the Secretariat was largely on the frontline and in control of the negotiating process. As a result, the credibility of the Commission as a whole was under-utilised. Had negotiation occurred through the Commissioners, the resultant terms of reference would have had greater credibility and been less subject to question later in the process. In the early stages of the WCD, the Commission intended to organise Programmatic Committees that would have allocated specific reviews to specific Commissioners. Under pressure of time and workload, this evolved into a looser structure in which Commissioners expressed an interest in particular thematics, but without a corresponding, defined set of responsibilities for their chosen areas. In hindsight, a more structured approach might have provided a better vehicle for Commissioner inputs.

Box 5.5

What is the "right" question? Economic, Financial, and Distributional Analysis

The experience of the Economic, Financial, and Distributional Analysis thematic review points to the importance of the research question in framing the final output. It also illustrates the negotiation that occurred over some thematic reviews.

The initial terms of reference called for a review of the capacities and limitations of cost-benefit methodologies. Even at this early stage, there were indications of discontent, mostly from civil society groups, with the phrasing of the question. The first draft of the review paper stuck narrowly to this limited scope. The Commissioners received it poorly, calling for more discussion of actual practice. Civil society reviewers argued for empirical evidence of the accuracy of cost-benefit analyses on dams. Consequently, a practitioner of cost-benefit analysis was asked to prepare a second draft with more discussion of practice. This paper, too, did not meet with the approval of the Commissioners, one of whom dismissed it as "half-naked!"[a] Academic reviewers declared this version a step back from the state of knowledge on the topic, and civil society groups argued it lacked a comparison of performance and projections. Finally, a Secretariat staff member prepared a third paper, drawing on earlier drafts and on submissions to provide empirical detail.

The reviewers' reactions were only in part dictated by the quality of the various drafts. Also at stake was the emphasis of the review and its implications for the various stakeholder positions. Thus, a narrow theoretical analysis of cost-benefit analysis would have illustrated the potential of the technique, but would have failed to reveal flaws in implementation. An assessment of practice alone would have highlighted procedural flaws, but would not have provided details on how past dams have performed. An exclusive focus on empirical experience would have allowed an assessment of past experience, but not shed light on whether the problem lies in flawed implementation or deeper problems with the approach. These alternatives were preferred to different extents by the various stakeholders, each of whom tried to advance their interests in the design of the thematic review.

[a] Interview with Commissioner, April 2000.

Source: Based on a review of the Secretariat's archival material by Luna Ranjit, WRI.

Box 5.6

Negotiation over content: the Social Impact thematic review

The thematic review on Social Impact of Large Dams: Equity and Distributional Issues required explicit negotiation among stakeholders. This thematic was one of three in the social issues category; the other two focussed on indigenous peoples and displacement. From the start, a problem of scope plagued this thematic. Initially, the paper was designed by the Secretariat to fill gaps left by the other thematics, notably downstream social impacts and gender impacts of dams. However, the Commissioners expressed their dissatisfaction with the patchy framework for the paper and sought an expanded paper that would address two concerns. First, some Commissioners argued that there was inadequate attention to the benefits of dams. Second, others suggested that the issue was not simply one of aggregate costs and benefits, but their distribution. Hence, they sought to locate social impacts within a framework of equity analysis.

Reviewers picked up these themes in a more partisan manner. Industry groups and irrigation specialists charged that the Social Impacts thematic paper did not address the social benefits of large dams at all and was flagrantly biased. This view had at least some support on the Commission; one Commissioner bluntly stated that the report's authors were "too far to the left."[a]

Resolution was sought in a special meeting convened by the Secretariat in London for reviewers to work through the outstanding issues. By all accounts, this was a spirited, but productive, meeting. Much of the discussion focussed on an effort to elaborate a framework for an equity analysis of the social impacts of dams. At this point, the scope of the paper had expanded considerably and posed a challenge of synthesis. As a result, the final document was segmented into sections on equity, downstream impacts, and gender impacts. The framework on equity incorporated into the paper was perceived to be a step forward by those who had espoused this argument. Proponents of more attention to the benefits of dams were, however, less satisfied, and this issue continued to be contentious through the life of the Commission.

How do we view such a process? As we have seen, reviews of controversial topics as part of a multi-stakeholder process will invariably entail a measure of negotiation. Indeed, the strength of the process lies in the opportunities for all sides to put forward their views and strive for common ground. As with other thematic reviews, the scope of the initial framing question proved to be central to the product's acceptability to stakeholders.

[a] Interview with Commissioner, April 2000.

Source: Based on Secretariat documentation and communication on the Social Impacts thematic; Interviews with Commissioner, 26 February 2001; Secretariat staff, 26 February 2001; and consultant, 2 March 2001.

Public Submissions: A Participation Channel that Needs Resources

The 970 submissions made to the WCD by a range of stakeholders from around the world were arguably the most significant casualties of the time pressure the WCD faced. The submissions were solicited both through the regional consultations and independently, and were intended to be a means of opening up the knowledge generation process. However, the mechanism for ensuring that submissions were incorporated into the thematic reviews was inadequate for two reasons. First, the timing of submissions and thematic reviews was not synchronised. The Commission was still receiving submissions long after drafts of the thematic reviews were completed.[62] Second, consultants were not always amenable to incorporating submissions into their work. Although Secretariat interns were allotted the task of processing the submissions for easy inclusion into thematics, it is not clear this effort met with success. Indeed, none of the consultants contacted in the course of this assessment acknowledged the receipt of submissions to incorporate into their work.

Public submissions were inadequately incorporated into the Commission's work.

In addition, the perspective of at least some of the submission writers and the consultants illustrated a clash in perspectives over the scientific nature of knowledge. The authors of the Ecosystem thematic, for example, sought quantitative knowledge rather than anecdotal inputs.[63] Thus, the Commission's efforts at democratisation of knowledge were affected not only by time pressures but also by opinions in the Secretariat and consultant body over the credibility of different forms of knowledge.

The Power of the Pen: Selection of the Research Team

As in the case studies, the mix of consultants chosen to implement the thematic reviews affected the WCD's ability to engage a wide range of stakeholders. The Secretariat recognised the importance of consultant selection from an early stage and called for a range of disciplinary approaches, institutional backgrounds, and political perspectives among consultants. In addition, the selection process placed a premium on geographic, racial, and gender balance. In general, the Secretariat strove for impartial writers. Where thematic studies were the outcome of a panel or task force, the Secretariat attempted a balance of perspectives.[64] As one Secretariat member put it, the consultant selection process mirrored the efforts to establish balance in the Secretariat itself. Putting together a consultant team was like establishing "mini-secretariats."[65]

There proved to be a number of barriers in translating policy into practice. First, the short timeframe of the Commission's work placed limits on whom the Commission could call upon. For example, researchers with full-time positions were unable to commit to the deadlines demanded by the Commission's workplan. The result was a smaller available pool of contributors, limited to short-term consultants, which potentially compromised the quality of the knowledge base. Moreover, in the eyes of some civil society Forum members, short-term consultants tend to operate within a mainstream development framework associated with large development bureaucracies, because the bulk of their work is for these bureaucracies.[66] Civil society members felt this perspective was carried over to consultants' work for the WCD and systematically skewed it toward a mainstream orientation.[67]

Given the short time frame, the pool of available consultants was small.

An analysis of consultants' backgrounds partially supports the view that they had a mainstream orientation. *(See Figure 5.1.)* The single largest category of thematic review writers was academic/research (36 percent), and the second largest was consulting (34 percent). However, these are slippery categories and only imperfectly reflect the issue at hand—a perception of mainstream mindset.[68] Moreover, some Secretariat members found that consultants gravitated toward standard research models based on their work for international agencies, and that it was a challenge to force consultants "out of their own little world."[69] In

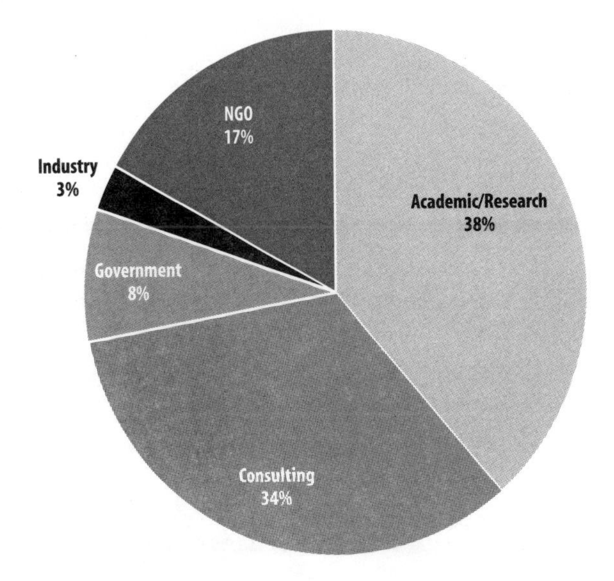

Figure 5.1 Consultant background (thematic review)

Based on information available from 84% of 120 total thematic review consultants.

Source: Data provided by the Secretariat.

summary, the consultants dissatisfied civil society groups engaged in the process and, at the time, tempered many groups' support for the process. Consultants' lack of familiarity with a broad, multi-stakeholder approach imposed an additional supervisory burden on Secretariat staff. Although the WCD's aspirations for consultant use provide a good standard for future processes, its experience demonstrates the practical hurdles involved.

Second, with the Chairperson and Vice-Chairperson of the WCD both from the South, the Commission was finely attuned to the need for adequate representation from Southern countries on the research teams. Yet, this proved difficult to put into practice in the selection process. Researchers from the South were often national or regional experts who would have been hard pressed to conduct a global review. The requirement that work be conducted in English posed a further challenge to recruitment. Thus, 56 percent of the consultants used for the thematic review were from North America and Europe.[70] *(See Figure 5.2.)* Moreover, Commissioners were concerned that lead writers were disproportionately chosen from the North, and particularly from the English-speaking countries.[71]

Women were similarly under-represented in the consultant pool and hence in weaving their

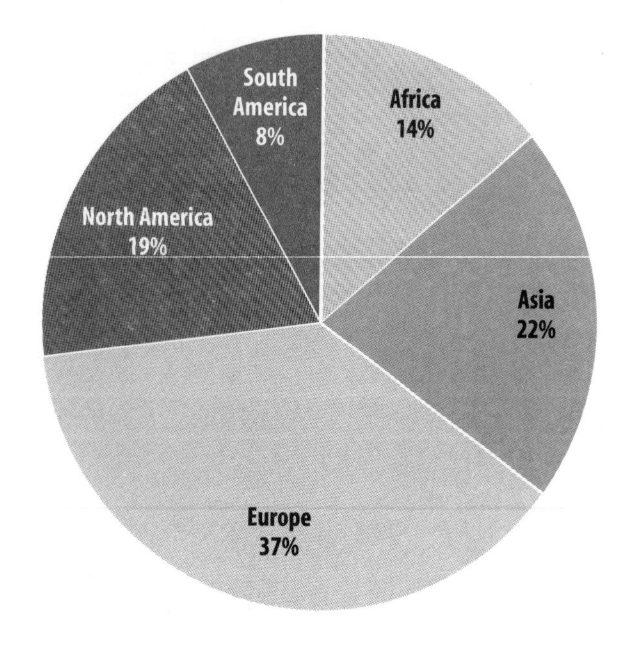

Figure 5.2 Consultant nationality (thematic review)

Based on information available from 61% of 120 total thematic review consultants.

Source: Data provided by the Secretariat.

perspectives into the WCD's formal knowledge base. A cross-cutting analysis of the Commission's consultants shows the low percentage of women consultants—only 25 percent of the total.[72] This is surprising given the Commission's relative emphasis on understanding the environmental and social problems associated with dams, rather than on the technical aspects of dam engineering, a more heavily male-dominated field. The Commission may have been stymied by discrimination (including lack of qualified female analysts) in the particular countries where it chose to work.[73]

Cross-cutting Issues

Assistant, Professor, Editor, and Referee: The Roles of the Secretariat

The work of the WCD, opined a Commissioner, is like cooking vegetables. In this process, the Secretariat and the Commissioners have distinct roles. "Once the vegetables are bought, cleaned, and cut into the necessary pieces, the role of the cook begins. The Secretariat is not the cook. We [the Commission] will decide the taste, the flavour, the aroma and the temperature ... "[74] This metaphor appropriately captures the expected division of labour between the Commissioners and the Secretariat. It also, however, captures the ambiguity of the Secretariat's role. The Secretariat was a helper, but in its choice and preparation of the raw materials had considerable influence over the final dish.

In the design of the WCD, Secretariat staff members were the first to phrase the questions and themes that the Commissioners discussed. Secretariat staff wrote overarching background papers, and terms of reference for consultant papers. The Secretariat's work was not accepted unquestioningly. On several occasions, such as the social thematic and the financial thematic, the Commission exercised its right to send the draft back with instructions for a complete re-write. Although the Commissioners were highly dedicated, most also had other ongoing commitments and were unable to devote all their time to the Commission. In the context of limited Commissioner time and attention, the role of the Secretariat in framing issues gave them, as one Commissioner put it, the "power of the professor."[75]

In addition to framing debates, the Secretariat also exercised the power of the editor. The Secretariat bore the enormous burden of synthesising large amounts of carefully worded, and, in some cases, negotiated, text into brief summaries for the Commissioners' consumption. This is not to suggest that Secretariat staff consciously filtered information for the Commissioners' consumption; however, the summarising process inevitably required Secretariat staff to exercise their judgement of the relative weight of arguments and issues raised in thematic papers. Commissioners were acutely aware of the Secretariat's filtering role, and in some cases, sought to read unedited documents.[76]

The Secretariat was also a referee. In the negotiation process through which case studies and thematic reviews were defined, written, and revised, the Secretariat was the gatekeeper of the Commission's neutrality. This was a challenging task. Secretariat members had to ensure that all sides were represented in the review process and had to establish and defend the line between input and undue influence.

Finally, the Secretariat's influence was amplified by their pre-eminent role in consultant selection. As one might reasonably expect, the primary sources of candidate consultants were the staff members' own professional networks.[77] Commissioners and Forum members' suggestions supplemented Secretariat selections, but primary control rested with the Secretariat. Indeed, some Commissioners made it clear that consultant selection was not their job and even viewed the active role of their colleagues as interference in the work of the Secretariat.[78]

The Importance of Accessibility: Language and the Internet

Although the country meetings provided communities with an opportunity for direct participation in the WCD process, these meetings also revealed one of the greatest stumbling blocks to grassroots participation: inadequate translation of documents from English. Language problems mounted a common challenge to full stakeholder participation in all case studies (except the United States) and limited the scope for meaningful local input. These problems were of sufficient magnitude to undermine the Commission's legitimacy with civil society groups in some case study countries. For example, an Urdu language summary of the draft Tarbela case study only became available to participants on the morning of the second stakeholder meeting, causing discomfort among local civil society participants.[79]

Language problems posed a challenge to full stakeholder participation.

The almost exclusive use of English to conduct Commission business, including negotiating terms of reference for thematic papers, discussing meeting agendas, and so forth, left some non-English speakers feeling that their participation was compromised.

> "Communication between MAB [Brazil's Movement of Dam-affected People] and the WCD's Commissioners and Secretariat was mainly through e-mail and postal mail because these are mediums in which one can think, prepare, and ask someone else to translate before sending the final message. This long process ensured that MAB's participation frequently lagged behind. MAB tried as best they could to keep up with the WCD's timing, but it was nearly impossible.
>
> [There was] another problem characterised by the movement's leaders as 'second hand information.' They were continuously reading documents translated by different people that could be excellent partners, such as NGO activists and academics, but were not dam-affected themselves. 'We were constantly limited in the process because we always had to analyse information from someone else's point of view.'"[80]

The Secretariat argued that these difficulties were both a function of time and budget. To keep their timeline on track, they were not willing to translate long case study drafts into national languages.[81] However, it could be argued that translations should be built in as an integral part of the time line and scope of work from the start. Given the broader historical tendency for global public policymaking processes to be elitist in nature and the thrust of the WCD toward greater inclusion, the problems posed by language issues partly undermined the Commission's larger effort.

For the process as a whole, language barriers, compounded with reliance on the Internet for communications, posed a double bind for participants in the South. On the one hand, electronic mail (e-mail) technology contributed greatly to the participation of certain groups and individuals from Southern countries in the Commission's work. The Southern members of the Commission itself were able to communicate quickly and efficiently with the Secretariat and fellow Commissioners by e-mail. Secretariat members attributed much of their responsiveness to Commissioner and Forum members' concerns—spontaneously and across multiple time zones—to their e-mail connectivity.[82] The technology also helped the Secretariat supervise consultants and elicit results within the ambitious timeframe mandated. On the other hand, the Internet sped up the process so much that, as the Brazilian example above illustrates, groups in the South were challenged to keep up. This experience joins a body of analytic material on multi-stakeholder processes documenting the dangers of relying too much—or exclusively—on Internet communication because of disparities in access.[83]

Conclusions

The investigative process inevitably affects public perceptions of legitimacy in multi-stakeholder

Strategies for building credible multi-stakeholder processes

- Adopt an explicit commitment to good governance in order to create the political space for engagement of diverse voices—this diversity is what demonstrates the value-added of the multi-stakeholder process.
- Adopt a work programme that allows stakeholders to propose diverse approaches and measures in order to foster inclusion.
- Budget sufficient time for concerned interest groups to become informed about the work programme and participate in its shaping.
- Choose local consultants rather than international consultants to engage in fact-finding wherever possible, but check their political acceptability with a range of local stakeholders to avoid disenfranchisement.
- Recognise that perceptions of the relative roles of the commission and secretariat in directing the process can affect stakeholders' trust. The commission should identify and take clear control over decisions that are controversial in stakeholders' eyes.
- Budget sufficient time and money to translate framing documents and synthesis outputs into other languages, especially for countries where significant consultation and data gathering is taking place.
- Do not allow the speed of Internet communication to speed up the pace of the work programme beyond the ability of non-connected, non-English speakers to participate.

processes that study past practice as the basis for forward-looking recommendations. Stakeholders ask whether the process is balanced, inclusive, transparent, and fair; and their reception of the recommendations depends on the answer being "yes."

Several major aspects of the WCD's work programme affected its credibility. First, the explicit effort to design good governance principles into all components of the work programme earned stakeholder trust. This commitment helped to engage interest groups in negotiation over the framing and composition of the work programme. In other words, the WCD's efforts to cast itself as an honest broker and open listener in gathering knowledge about large dams provided the preconditions for the broader stakeholder involvement that followed.

Second, having established that political space, the work programme became a platform for heated contention among interest groups, which vied for influence in shaping the way issues were framed. Did a dam damage an ecosystem irreparably? Did it create new, viable ecosystems of its own? It was over the phrasing and emphasis of such questions that interest groups pressured the WCD—and primarily the Secretariat as mediators—in the course of knowledge gathering.

This ability to influence the work programme was empowering for interest groups that had access to the WCD's work through their networks (contacts with either Secretariat or Commission members), their English language ability, and their access to telecommunications technology. For concerned stakeholders with more limited access to the Internet or English language, events moved too fast for their meaningful participation and they were reliant on information filtered through secondary sources. The process was disempowering for them. Even stakeholders who wanted to participate in the shaping of the work programme and had easy access to Commission publications found the time for digesting material and providing input too short. Their complaints led to time-consuming negotiations later.

The lesson for future processes is that ample time must be budgeted for informing stakeholder groups of the process' aims. For political acceptability, a core group of stakeholders (such as, in the WCD's case, the advisory Forum members) must have the chance to comment on the direction of the work programme. This accomplishes two benefits: groups can then mobilise their own resources to contribute to the work programme; and they can negotiate contested concepts early on, which reduces the need for expensive course corrections later.

Another lesson, arising from the overall negotiation of the work programme as well as the individual country and river basin meetings, is the need for more document translation. Although expensive, time and money for translations and interpretations should form an integral part of budgets and workplans in future processes of this kind. Because it was not practical to translate multiple drafts of working papers for stakeholder dissemination, a reasonable standard may be to translate essential framing documents and interim products into major world languages.

A third major area for consideration in future multi-stakeholder processes relates to the difficult interface of global forums with individual country politics. Future processes will face the same trade-offs as the WCD did, whereby the WCD sought to create the space for multi-stakeholder dialogue in each of its case studies. However, this aspiration limited the choice of countries where it could work. In the semi-authoritarian countries where the WCD chose to work (based on the dams population), civil society input was curtailed. Participants in future processes will have to choose between promoting dialogue and gathering data in countries with open political systems where inclusiveness will not be a challenge versus pushing the envelope on inclusivity in less open political regimes.

The explicit effort to build good governance in the work programme earned stakeholder trust.

Debates over which kinds of data belong in the knowledge base will play a prominent role in any future multi-stakeholder process with a serious fact-finding component. The knowledge gathering process often requires several contributors from one place to provide a range of perspectives and, therefore, ensure credibility. Political tensions around access to and privilege over scientific data posed major problems to both the comprehensiveness of the WCD's knowledge base and its acceptability to various stakeholders. In some cases, project owners were simply unwilling to share data because of security and other concerns. In other cases, the WCD chose local consultants who had access to official data but such consultants typically alienated civil society groups by neglecting qualitative and experiential forms of knowledge.

Finally, future processes will need a carefully negotiated division of labour between the commission and secretariat that takes fully into account interest group sensibilities about bias and dissipates tension with the fullest transparency possible. It may be possible for future commissions to identify the management issues that are especially sensitive to stakeholders and to have a greater say in them. In terms of the WCD, the work programme was so ambitious that the Commission had to delegate most of the direct fact-finding to the Secretariat and consultants, while it took the role of weighing the evidence and reaching final conclusions. Given the contested history of the Commission's formation, based on negotiation by interest groups, this delegation to the Secretariat and consultants (who were *not* chosen through such a shared process) concerned stakeholders who feared the introduction of bias. The neutrality of the Secretariat and consultant body remained a hot issue throughout the process, especially for NGO advocates.

Endnotes

1. WCD Secretariat, "Draft Briefing Note 3: Some Thoughts on WCD Secretariat Organisation and Structure." Prepared for the first Commission meeting, May 1997.
2. World Bank/IUCN, *Large Dams: Learning from the Past, Looking to the Future* (Gland: IUCN, 1997), p. 11.
3. Interview with Gland workshop organiser, 1 February 2001.
4. Sourced from internal Secretariat documents, "Draft Briefing Note 2, Some Considerations in Developing the Framework for the World Commission on Dams' Work Programme," May 1998.
5. Secretariat presentation to the WCD Forum, 5-6 April 2000, Cape Town.
6. Sourced from internal Secretariat documents, "Draft Briefing Note 2, Some Considerations in Developing the Framework for the World Commission on Dams' Work Programme," May 1998; "World Commission on Dams Strategy and Objectives June 1998-June 2000," August 1998.
7. Work Programme of the World Commission on Dams, February 1999.
8. Patrick McCully, "A Critique of 'The World Bank's Experience With Large Dams: A Preliminary Review of Impacts,'" April 1997. Online at: www.irn.org/programs/finance/critique.shtml (28 September 2001).
9. Interview with Commissioner, April 2000.
10. Medha Patkar, speech to NGOs, December 1999.
11. Personal communication from Secretariat staff, 3 July 2001.
12. Interview with senior Secretariat staff, April 2000.
13. Interview with senior Secretariat staff, April 2000.
14. Early iteration of the workplan, thematic reviews, internal Secretariat document, August 1998.
15. Interviews with senior Secretariat staff, 6 November 2000.
16. World Bank/IUCN, 1997.
17. Interview with WCD webmaster, November 2000.
18. Interview with World Bank official, 22 January 2000.
19. Interview with senior Secretariat staff, 6 November 2000.
20. Interview with World Bank official, 22 January 2001.
21. Interviews with Secretariat staff, December 1999 and April 2000.
22. WCD website, www.dams.org/studies/ (28 September 2001). A comprehensive official version of the case study methodology is given on pages 350-352 of the WCD's final report.
23. Interview with senior Secretariat staff, November 2000.
24. World Commission on Dams, *Dams and Development: A New Framework for Decision-Making* (London: Earthscan, 2000), p. 351. Also, interview with and internal documentation from senior Secretariat staff, 6 November 2000.
25. Interview with senior Secretariat staff, 6 November 2000.
26. Interview with senior Secretariat staff, 6 November 2000. The Ataturk dam on the Euphrates River has been the source of tension between Turkey and its neighbours. The World Bank refused to fund Turkey's dams on the Tigris and Euphrates rivers due to water rights controversies in the region.
27. Personal communication with Flávia Braga Vieira, January 2001; also Flávia Braga Vieira, "Brazil's Dam Affected People's Movement and the World Commission on Dams." Background paper prepared for the WCD Assessment, January 2001, p. 10.
28. Comments by Secretariat staff during focus group session, 3 November 2000.
29. Comments of senior Secretariat staff to the WCD Forum, 5 April 2000.
30. For example, public plea by dam opponent at April 2000 Forum meeting for Commission to scrutinise Narmada Valley dams more closely. See also Braga Vieira, 2001, p. 8.
31. Public correspondence of Achim Steiner, WCD Secretary General to Jacques LeCornu, ICOLD Secretary-General, 18 April 2000. Online at: http://genepi.louis-jean.com/cigb/steiner.htm (28 September 2001). Steiner wrote: "We particularly regret the misinformation that has led the Ministry of Water Resources to assume that the WCD was going to produce a final report about the Three Gorges project. The WCD is neither a Commission about the Three Gorges Project nor is it a Commission about Chinese dams. Our work programme is intended to assemble as much information as possible about dams that would enable the Commission to develop its own, independent understanding of the key issues and lessons learnt. In that context we attach great value to learning about China's approach to dams in the context of water and energy resources management."
32. World Commission on Dams, 2000, p. 351.
33. E-mail correpondence with staff of China human rights organisation, 30 March 2001. Interviews with participants of East and Southeast Asia consultation, Hanoi, 25–27 February 2000.
34. Work Programme of the World Commission on Dams, February 1999.
35. World Commission on Dams, 2000, p. 350.
36. World Commission on Dams, 2000, p. 350.
37. WCD Secretariat, "WCD Criteria for Selection of Consultants," draft, 14 October 1998.
38. Tundu Lissu, attendance at the second stakeholder meeting for the Grand Coulee Case Study, January 2000.
39. Interview with senior Secretariat staff, 6 November 2000; interview with Secretariat staff, 12 February 2001.
40. See Pak Mun Dam case study on www.dams.org/studies/th/ (28 September 2001).
41. Interview with Secretariat staff, 8 December 1999.
42. Interview with private sector representative, 26 February 2001.
43. Interview with senior Secretariat staff, 6 November 2000.
44. They also organised the Tucurui stakeholder meeting to ensure the participation of affected people, according to a senior Secretariat staff member in a November 2000 interview. Melchisedeck Lutema, attendance at the second stakeholder meeting for the Kariba Dam case study, 21-22 February 2000.

45. Elena Petkova, attendance at the second stakeholder meeting for the Aslantas Dam case study, January 2000. The Secretariat briefed Turkish environmental NGOs on the Aslantas case study in the scoping stage, but the NGOs were stretched thin by other commitments and declined to get involved, according to a personal communication from Secretariat staff, August 2001.
46. Interview with Commissioner, 6 April 2000.
47. Eventually, the survey covered only 125 dams because of incomplete survey forms.
48. Interview with Secretariat staff, 4 November 2000.
49. World Commission on Dams, 2000, p. 353.
50. World Commission on Dams, 2000, p. 354.
51. E-mail correspondence from South African NGO to World Resources Institute, May-June 2000.
52. Interview with Secretariat staff, 4 November 2000.
53. Interview with Secretariat staff, 4 November 2000.
54. Interview with senior ICID official, 6 April 2000.
55. Interview with Secretariat staff, 4 November 2000.
56. Interview with Secretariat staff, 4 November 2000.
57. World Commission on Dams, "Strategy and Objectives: June 1998-June 2000," Cape Town.
58. Interview with industry representative on the Forum, 6 April 2000.
59. Ramananda Wangkheirakpam, "Assessment of the WCD Processes on Thematic V.4." Background Paper prepared for the WCD Assessment.
60. Interview with Secretariat staff, 17 August 2000.
61. Interview with Secretariat staff, 17 August 2000.
62. The exception that proves the case is the "Financial and Distributional Analysis," which did manage to incorporate submissions only because the thematic review was substantially delayed. Interview with Secretariat staff, 8 April 2000.
63. Interviews with Secretariat staff, 3 November 2000 and 6 November, 2000.
64. WCD Secretariat, "WCD Criteria for Selection of Consultants," draft, 14 October 1998.
65. Interview with Secretariat staff, 13 December 2000.
66. This is a difficult charge to fully substantiate since the categories of consultant and researcher are blurred in practice. Thus, 43 percent of WCD consultants were identified by the Secretariat with academic/research, while 28 percent were termed consultants.
67. Interview with civil society representative on the Forum.
68. This was based on a categorisation of WCD consultants by Secretariat staff rather than on a self-reporting basis. These numbers should be read only as a rough indication since categories such as "consulting" and "academic/research" shade into each other at the margins.
69. Interview with Secretariat staff, 17 August 2000.
70. The Secretariat maintained records for consultants based on nationality and "place of work." Of the 65 percent who reported their nationality, 56 percent were European and North American. Of the 91 percent that reported place of work, 63 percent were European or North American.
71. Interview with Commissioner, December, 1998.
72. Based on an analysis of consultant data provided to the assessment team by the Secretariat.
73. Interview with Secretariat staff, April 2000.
74. Interview with Commissioner, August 2000.
75. Interview with Commissioner, 7 April 2000.
76. Interviews with Commissioners, 7 April 2000, November 2000.
77. Interviews with Secretariat staff members, 17 August 2000, April 2000, 7 April 2000.
78. Interview with Commissioner, 7 April 2000.
79. Gopal Siwakoti, attendance at the second stakeholder meeting of the Tarbela case study, January 2000; interview with Commissioner, April 2000.
80. Braga Vieira, 2001, p. 19.
81. Interview with senior Secretariat staff, 6 November 2000.
82. Interview with senior Secretariat staff, 6 November 2000.
83. Anthony Dorcey, *Institutional Design and Operational Modalities for the Proposed Large Dams Commission*, Stockholm Draft, 6 August 1997 (mimeo): Personal communication with UNED Forum, 3 August 2001.

Chapter 6

Implementing the Work Programme: Consultations and Outreach

Public consultations have increasingly become a norm for development policy processes at the regional and global levels. Because the Brundtland Commission gave public consultations higher visibility, decision-makers have viewed them as ways of raising awareness of forthcoming policies, providing a sounding board for policies' acceptability, and strengthening their content. In the recent history of global commissions, public consultations have played an important role in helping commissioners to define their problem statement and sustain wider public engagement. Multilateral development banks have increasingly built public consultations into policy processes.

For the World Commission on Dams, the regional consultations served both a fact-finding purpose and a way of raising the profile of the Commission and its work around the world. In a symbolic sense, the regional consultations were intended to portray the WCD as an open, listening commission and to make Commissioners more accessible to people. In this way, the decision to hold regional consultations grew directly from the Commission's commitment to inclusiveness and transparency and its commitment to project an appropriately unbiased image. However, raising the public profile also brought the Commission under greater scrutiny and increased the risk of alienating interest groups, if consultations went wrong.

This chapter examines the Commission's success in fulfilling the multiple objectives of the public consultations: Did the Commission gather diverse viewpoints from its consultations? Did it raise the profile of its work with concerned stakeholders? And most important, what aspects of the consultations strengthened or undermined the Commission's broader legitimacy?

A Listening Commission: The Regional Consultations

The Commission undertook four regional consultations: in South Asia, Latin America, Africa and the Middle East, and East and Southeast Asia. *(See Box 6.1.)* In total, 1,400 individuals from 59 countries took part in the regional consultations.[1] The Commission also held 20 country and river basin-level consultations to discuss its case studies.

Box 6.1

WCD regional consultations

- South Asia—Colombo, Sri Lanka, 10–11 December 1998
- Latin America—São Paulo, Brazil, 12–13 August 1999
- Africa / Middle East—Cairo, Egypt, 8–9 December 1999
- East and Southeast Asia—Hanoi, Vietnam, 26–27 February 2000

Source: WCD website, www.dams.org/consultations/ (28 September 2001).

Given limited funds and time, the Commission decided to focus on listening to stakeholders in world regions where dam building was high on the agenda for future development, i.e., in developing countries.[2] By 1998, Europe and North America had some of the longest experience in dam building, but they had largely exploited their hydro potential. Stakeholders widely accepted the emphasis on Southern regions: Not only did those regions face the greatest challenges in water and energy supply, but their citizens also had the greatest difficulty in accessing international policy fora. It was fitting that the Commission should come to them.[3]

The consultations provided legitimacy for the Commission's final report.

When it came to marketing the Commission's final report, the consultations served as a strong public relations tool for legitimising the product. Given the historical precedent of and increasing expectation for consultation in local, national, and international development processes, the absence of consultations would likely have undermined the Commission's credibility, particularly with civil society groups. Such mechanisms were incorporated in concurrent environment and development processes, such as the World Water Vision process. *(See Box 6.2.)*

The Sensitive Interface of Global Process and Local Politics

The WCD's first consultation, planned for the South Asia region and to be held in India, ran headlong into the complexities of domestic politics. Instead of an opportunity for the WCD to model how it would listen to different sides of the dams story, the event turned into a lesson on the perils of national and local politics for the perceived independence of a global commission.

The Commission chose Bhopal, in the Indian state of Madhya Pradesh, as its meeting site. The neighbouring state of Gujarat is intended to be the prime beneficiary of water and power from the huge and controversial Sardar Sarovar Project on the Narmada River. The Narmada River, for which more than 3,000 dams have been proposed, runs through both states. Over time, the social movement opposing the project has built up a remarkable national and global alliance of support groups. Major donors have withdrawn from the project in light of its serious social, environmental, and economic impacts, including the World Bank whose Morse Commission sounded the alarm following a thorough inquiry. This history and context made the Gujarat government particularly sensitive to international intervention.

The WCD's field trip to the Narmada Valley construction and protest sites, planned by the activist Indian Commissioner Medha Patkar, caused great concern to Gujarati officials. The Government of India already suspected the WCD of an anti-dam bias, as recounted in Chapter 4, simply because Ms. Patkar was on the Commission. When news of the proposed siting of the consultation in Bhopal and the Commissioners' field trip to the Valley filtered out, the officials felt their worst fears about international intervention had been confirmed. The Gujarat assembly pressured the national government to withdraw its permission for the WCD meeting. The national government did so, only four days before the scheduled meeting date. The state government called the proposed WCD visit "an invasion by developed nations on under-developed countries."[4] The Indian press carried negative reports of the WCD. Of the many columns that appeared, one went so far as to dub the WCD a "fraud commission."[5] Another characterised the Commission as "conceived last year by a group of about 40 people from various countries to launch a campaign against large dams."[6]

The WCD's proposed field trip to the Narmada Valley concerned Indian government officials.

This difficult beginning tarnished the Commission's reputation with external audiences and led to much soul-searching within. It taught the Commissioners and Secretariat a lesson they would never forget: the consultations of a global body can be highly sensitive in certain local and national contexts. The Commission's substitute regional consultation for South Asia was held in Colombo, Sri Lanka, in December 1998. Commissioners and Secretariat staff lauded the event for its success in bringing opposing sides of the debate to the table for constructive exchange. "I never imagined seeing such democracy at work," said Medha Patkar of the diverse participation at the Colombo meeting.[7] The Commission continued its regional consultation events in São Paulo, Cairo, and Hanoi.

Meeting Participation under Restrictive Political Regimes

The Commission chose country venues for its regional consultations based primarily on practical

Box 6.2

The World Water Vision process

The WCD process took place at the same time as an evaluation of water management at the global level: the "Long Term Vision for Water, Life and Environment in the 21st Century," or World Water Vision process (WWV, 1997-2000). The objective of WWV was to formulate a shared vision for how to mitigate the forthcoming global challenge of water scarcity. The exercise was based on regional and sectoral visions produced by stakeholders through a series of consultations. A complementary World Commission on Water was formed in July 1998 chaired by Ismail Serageldin, then a Vice President at the World Bank, to issue an independent report based on the visioning exercise.[a]

According to the Vision report, the WWV's consultative process involved "authorities and ordinary people, water experts and environmentalists, government officials and private sector participants, academics and NGOs." The organisers estimate that at least 15,000 people were directly involved in drafting Vision documents for specific regions and sectors through these consultations.[b] Although the WWV reached a greater number of people through its consultative process than the WCD did, the WWV provided a platform for a narrower range of stakeholders to express their views than did the WCD.

The sectoral consultations were organised around four principal themes: Water for People; Water for Food and Rural Development; Water and Nature; and Water in Rivers. Major water experts and water-related interest groups, such as the International Committee for Irrigation and Drainage (ICID) and members of the Consultative Group for International Agricultural Research (CGIAR) system, organised these consultations and wrote the resulting "Vision" documents.

The regional consultations were initiated by the regional committees of the Global Water Partnership, a pre-existing network representing "eminent expertise within each region on water resources management."[c] They took place in all major continents, as well as in major river basins or riverine systems, such as the Nile and Aral Sea Basins. Notably, the World Water Vision's regional consultations were decentralised: "A key part of the "contract" with the groups doing the consultations was that they would be free to identify the issues of concern to them and draw their own conclusions."[d] This structure was nearly opposite to the WCD's, in which consultations took the form of hearings and were tightly organised under the central control of the Secretariat staff. The promise of the WWV model was that stakeholders might achieve a more collaborative interaction and greater ownership through direct participation. The risk of the model was that consultations might be dominated by the actors with greatest power and authority, the water "establishment."

Widespread accounts of the consultations, including accounts by the WWV Secretariat itself, indicate that governmental and quasi-governmental water agencies did play a dominant role.[e] There was a greater effort to incorporate women's groups and NGOs during the latter half of the process after these groups complained that they felt excluded.[f] And the sectoral consultation on Water for People, co-ordinated by the Collaborative Council on Water Supply and Sanitation, was notable for its attempt at a "bottom-up approach."[g] But the Vision's organisers concede that overall, the process was not as inclusive of grassroots and civil society inputs as they had hoped.[h]

The WWV process resulted in a global report that painted future scenarios for water use and water scarcity in broad brushstrokes and provided general recommendations for averting an acute crisis. The sheer number of large and influential water agencies involved in the process ensured that it captured headlines when the Vision report was released in March 2000 in The Hague. This event, the Second World Water Forum, involved a ministerial meeting that was well attended by international agencies and NGOs.

Advocacy NGOs were highly critical of the WWV process, to the point that they issued an alternative vision for management of the world's water.[i] One NGO press release called the WWV consultations a "sham." It charged, "The process has been controlled from the start by a small group of aid agency and water multinational officials, mainly from the Global Water Partnership, World Water Council, World Bank and Suez-Lyonnaise des Eaux. The key conclusions of the WCW [World Commission on Water] report that there is a global water shortage crisis which can only be solved with a massive increase in private funding for water projects in developing countries, backed up with guarantees from the World Bank and other aid agencies was predetermined."[j]

---

a. World Water Council, World Water Vision: Making Water Everybody's Business (London: Earthscan, 2000). Online at: www.worldwatervision.org/reports.htm (28 September 2001).
b. World Water Council, World Water Vision: Making Water Everybody's Business (London: Earthscan, 2000). Online at: www.worldwatervision.org/reports.htm (28 September 2001).
c. Global Water Partnership website, www.gwpforum.org (28 September 2001).
d. W.J. Cosgrove and F.R. Rijsberman "The Making of the World Water Vision," March 2000, p. 5.
e. see list of consultations for the World Water Vision process, in the appendix of World Water Council, 2000. Online at: www.worldwatervision.org/Vision/Documents/Appendix.pdf (28 September 2001).
f. W.J. Cosgrove and F.R. Rijsberman "The Making of the World Water Vision," March 2000, p. 2.
g. W.J. Cosgrove and F.R. Rijsberman "The Making of the World Water Vision," March 2000, p. 2.
h. Personal communication with World Water Vision Management Unit staff, 10 May 2001.
i. Both Ends, People-Oriented River Basin Management: An NGO Vision (Netherlands, 2000).
j. Excerpted from the statement "Old Water in a New Bottle: World Water Vision is Chronically Short-sighted." Written by International Rivers Network (USA), International Committee on Dams, Rivers and People, and Both Ends (Netherlands) and endorsed by 16 non-governmental groups from Brazil, England, India, Nepal, Pakistan, Slovakia, South Africa, Switzerland, and Thailand, 17 March 2000.

considerations: which governments were willing to grant permission, which locations were well situated geographically, and so on. For instance, Egypt was a natural crossroads between Africa and the Middle East.[8] Vietnam provided a good opportunity to influence dam decision-making as it had so many dams on the drawing board.[9]

The host country's political regime affected stakeholder participation in Commission events.

The Commission also had to consider how a host country's political regime affected participants' ability to express themselves. For a Commission seeking to promote a new consensus on dams and to operate in an inclusive way, the choice of country venue posed important trade-offs. Countries with open democratic systems would facilitate participation of diverse stakeholder interests. Those with less of a democratic tradition would potentially constrain it. The Commission faced choices between mobilising diverse stakeholder participation in countries with open political regimes, and advancing norms of multi-stakeholder participation in countries that traditionally allowed less political space for dialogue. Yet, as WCD Chairperson Kader Asmal noted, closed decision-making processes were the very reason for the WCD's formation, and were the practices the WCD most sought to change. This observation argued for holding consultations where democratic process was weak, with the intent of advancing norms of public participation.[10]

Among the WCD's choices, the political environments in Sri Lanka and Brazil allowed for expression of diverse viewpoints in the dams debate. By contrast, in Egypt and Vietnam the number of independent civil society groups was limited.

In Egypt, the government had passed a law in 1998 requiring governmental intervention in the management decisions of non-governmental organisations.[11] Environmental NGOs, while regional and global in their vision for environmental protection, dared not question the equity and benefits-sharing aspects of large water projects[12] and this was reflected in the limited nature of the Egyptian presentations at the WCD event. *(See Box 6.3.)* In Vietnam, the government has long discouraged independent civil society organising. The only independent Vietnamese voices at the Hanoi consultation, as opposed to ministry views, came from staff of an international NGO.[13]

At the same time, both the Egyptian and Vietnamese consultations, particularly the Hanoi meeting, illustrated the WCD's ability to be a catalyst for greater openness. The WCD consultation took place at a time of opening in Vietnam, only two months after the national government passed a decree recognising non-governmental entities ("foundations") for the first time. The Commission required freedom from restrictions on travel by participants from around the region as a condition for holding the meeting in Hanoi, and

Box 6.3

The missing voice of Egyptian Nubians

The Aswan High Dam, completed in 1972, is a huge monument to centralised development. The dam was built to reduce Egypt's vulnerability to flood and drought and expand its irrigated agriculture. A total of 120,000 ethnic minority Nubians were displaced from their homelands in Egypt and Sudan during the dam's construction.[a] The displaced Nubian population on the Egyptian side of the border reached 50,000. The government never recognised some families' claims for compensation.[b]

At the World Commission on Dams' regional consultation in Cairo on 8-9 December 1999, a morning was dedicated to hearings about Egypt's large dams experience. These presentations showed only the official history. They covered the development benefits to Egypt of the Aswan High Dam and, to a limited degree, contentions over environmental impacts. In spite of outstanding issues around compensation to the affected Nubian community and long-term impacts of loss of cultural heritage, social justice and distributional issues around water in Egypt were not part of the agenda.[c] The missing voices of the Egyptian Nubians from the regional consultation demonstrate the compromises to full and frank dialogue that occur when a global Commission, seeking to establish democratic process and norms, chooses to operate in an authoritarian context.

---

[a] Personal communication with Commissioner, August 2001.

[b] Interview with Egyptian Nubians, 11 December 1999. The assistance of Yomna Kamel in interpreting from the Arabic is gratefully acknowledged.

[c] Interviews with Egyptian Nubians, Cairo, 7-10 December 1999.

there were, indeed, no reported restrictions. In some ways, the government's hosting of the WCD meeting demonstrated willingness to experiment with a more open development dialogue. In the words of a Vietnamese social science researcher: "The Vietnamese government has created favourable conditions for NGOs in Vietnam now and democracy is increasing quickly."[14] Likewise, some Egyptian government officials were impressed to hear an "unusually" diverse range of views at the WCD regional meeting in Cairo.[15] Although these consultations were constrained, they may well have promoted the value of multi-stakeholder consultation in policy formulation. Future commissions will have to weigh carefully the choice of country—if they are to ensure rich, productive dialogue—and clearly negotiate the terms of the consultation with the host country in order to advance norms of participation in closed societies.

Challenges of Outreach to a Broad Range of Stakeholders

The WCD's modes of outreach for its regional meetings relied heavily upon the personal and professional networks of Commission, Secretariat, and Forum members. The Secretariat issued announcements and calls for submissions for the regional meetings, which were disseminated predominantly through the Forum's networks.[16] For instance, the World Bank, IUCN, International Rivers Network, and the International Commission on Large Dams (ICOLD) formed major outreach nodes. The Secretariat also sent announcements to contacts in government and in the Secretariat staff's own professional networks.[17] Typically, the announcements were translated into the host country language (Hindi, Portuguese, Arabic, Vietnamese) to raise local awareness. The WCD announced forthcoming events in its English language quarterly newsletter *Dams* (circulation 2000), and it posted news of forthcoming events on its website.

Civil society organisations on the WCD Forum and their networks made extraordinary efforts to solicit community inputs to the process. NGO and social movement representatives phoned, faxed, e-mailed, and met with dozens of community organisations, particularly in Africa, South Asia, and Brazil, to encourage them to make submissions to the WCD and participate in regional meetings. In addition, they organised and produced reports of regional hearings in South Africa and Western Europe, to which Commissioners and Secretariat members were invited. The hearings were intended as formal inputs to the Commission's process. *(See Box 6.4.)*

> *Civil society groups made extraordinary efforts to solicit community input for the process.*

The mobilisation by the Brazilian Movement of Dam-affected People (MAB) was particularly effective.[18] MAB saw the Latin America Regional Consultation as a key opportunity for dam-affected people without easy access to phone, fax, and e-mail to contact the Commission directly and to demonstrate the strength of their movement. The leadership anticipated that this event would be tremendously meaningful for grassroots participation and that it could be "the time for MAB to print its own stamp on the WCD process."[19] As a result of their efforts, 600 Brazilian dam-affected people attended the regional consultation in São Paulo. Limits on space precluded all the participants from filling the conference hall at once, and participants were disappointed at the lack of an open forum for exchanging views. However, they reported being moved by the experiences from neighbouring countries and heartened by their success in opening up the event beyond government, business, and academic participation. In all, 40 percent of the presentations at the São Paulo consultation came from MAB and its NGO allies.

However, civil society resources were limited, and the WCD did not compensate civil society efforts at outreach. Moreover, as with the Commission and Secretariat, civil society networks were not comprehensive either. For instance, the groups in Southern Africa that mobilised significant community input for the African regional event *(see Box 6.4)* had few contacts in East Africa, where controversial dam-planning and building are underway. In this case, adding another Forum member from the under-represented region may have boosted the WCD's ability to mobilise participation.

In a successful effort to broaden participation, the WCD did provide funding for developing country participants to attend regional meetings, based

Box 6.4

NGO-organised Regional Consultations and Mobilisation Efforts

Mobilisation in Africa

In Southern Africa, the Environmental Monitoring Group (EMG) along with the Group for Environmental Monitoring and the Botswana office of the International Rivers Network organised the "Southern African Hearings for Communities Affected by Large Dams." The hearing was not an official WCD event, but it was directed explicitly at the Commission. The event was intended to provide affected communities with an opportunity to present their experiences, which these NGOs felt had not been adequately covered in the Commission's work. The Secretariat staff, Chairman, Secretary-General, and one Commissioner attended.[a]

Participants produced a "Final Declaration: Voices of Affected Communities" that decried the social, cultural, economic, and ecological dislocations caused by large dams in the region and called for authorities to address outstanding historical injustices.

Based on contacts made in this process, EMG encouraged community groups to make formal submissions to the Commission. In addition, EMG staff e-mailed, faxed, and phoned all their contacts in community groups and NGOs throughout Africa to encourage submissions to the WCD. This outreach bore fruit: as a result of their efforts, EMG mobilised about 35 submissions to the WCD. The Commission selected seven of those authors to present their views in person and gave the authors grants to travel to the Africa and Middle East regional consultation in Cairo. According to a displaced man from Kwa-Zulu Natal, the information and moral support provided by EMG made his journey possible.[b]

Given that the WCD Secretariat's outreach for the Africa and Middle East consultation was limited largely to contacts in their staff's professional networks, such as the World Bank, IUCN, and governmental communities, along with stakeholders who had e-mail access, EMG's efforts to undertake a broader mobilisation proved invaluable in diversifying participation.

Mobilisation in Europe[c]

In Europe, NGOs led by Friends of the Earth–Slovakia organised a hearing entitled "Dammed at Home, Damming Abroad" in January 2000. The hearing was supported by private foundations and was closely co-ordinated with the Commission's schedule. Only one Commissioner and two Secretariat members attended the event, which was taken as a slight to the organisers.

Participants presented case studies of European dams and overseas dams involving European financing and construction. They raised various concerns about the WCD process and produced a letter to Chairperson Kader Asmal requesting that a long list of NGO concerns about dams performance be addressed in the remainder of the WCD's work programme.[d]

Mobilisation in India

In India, the Narmada Bachao Andolan (NBA), the movement with which Commissioner Medha Patkar is associated, and several other NGOs held four public hearings on large dams. As a national convenor of the larger National Alliance of People's Movements in India, Ms. Patkar was connected to a formidable network of community organisers, grassroots development workers, dam-affected people, and NGOs.

The network organised hearings in the states of Maharashtra, West Bengal, and Uttar Pradesh that covered a range of issues around flood management and irrigation dams and their impacts.[e] The events generated submissions that were sent directly to the WCD. NGOs and peoples' movements in India also translated the WCD newsletter, Dams, into Hindi and disseminated it. The NBA also sent out a regular e-mail update to a wide cross-section of people in India and abroad. Additionally, the South Asian Network of Dams, Rivers and People (SANDRP), based in Delhi, maintained regular communication with different stakeholders and facilitated the participation of civil society groups in the WCD process.

a. "Once There Was a Community: Southern African Hearing for Communities Affected by Large Dams: Final Report." Compiled and edited by Noel Stott, Karen Sack, and Liane Greef, Environmental Monitoring Group, March 2000. b. Interview with dam-affected community representative presenting at the Cairo consultation, December 1999. c. Summarised from Elena Petkova, World Resources Institute, trip report on attendance at "Dammed at Home, Damming Abroad." d. The letter is reproduced publicly on the European Rivers Network website, www.rivernet.org/nconf99.htm (28 September 2001). e. E-mail communication by Commissioner to the WCD assessment team, 15 February 2001

upon a selection process. This effort levelled the playing field for different actors. However, observers (as opposed to presenters) at the regional consultations had to pay their own way; therefore, issues of economic inequality affected participation in the meetings. Civil society groups could not afford to stay at some of the more expensive meeting venues. This experience, combined with concerns around communications media and language discussed in Chapter 5, raises accessibility concerns for multi-stakeholder processes that seek credibility through broad inclusiveness.[20]

Political Balance in the Consultations

Consultations constituted, above all, a public face for the Commission. Most Commissioners attended the regional consultations. Importantly, the appearance of the Commission as a group promoted the image of political balance, thus strengthening broad legitimacy.[21]

By contrast, it was impossible for all the Commissioners to attend the various additional country consultations. Only one or two Commissioners and Secretariat members attended each of these. This situation increased the risk that stakeholders would receive a biased impression of the Commission's political bent.[22] At country meetings, Commissioners had to make a greater effort to explain the diversity of views and experience in their larger group.

Achieving political balance in the consultations was important for many reasons.

In terms of content, at the regional hearings the Secretariat made efforts to project a neutral image by selecting a range of political views and topics for presentation. Each event lasted only two days so it was not possible to be exhaustive; instead, the Commission sought to hear a balance of experiences from the region as a whole.

A core group of Forum members, mostly from advocacy NGOs, monitored political balance on the agenda and were quick to chide the Secretariat when they felt diversity had been sacrificed. For instance, the Secretariat initially passed over a prominent domestic critic to choose a Chinese government speaker and foreign consultant to present information about China's dams experience at the East Asia meeting. Civil society groups staged an uproar.[23] In this case, although the Secretariat had sought political balance in its presentations from the region as a whole, China was such a large and significant dam-building country that to allow only pro-dam voices from that country to speak appeared overly biased. This issue was later resolved by allowing the domestic critic to address a closed-door session of the Commissioners. Aside from this example, in most cases the WCD assembled a sufficiently diverse set of presentations to satisfy Forum members monitoring the process.

The effort to create balanced presentations at the regional meetings was important for both Commissioners and participants. The Secretariat processed most of the numerous papers, studies, and submissions sent to the Commission. For Commissioners, the consultations provided a rare opportunity to receive information directly. Presentations by representatives of dam-affected communities about their experiences appear to have had a more profound effect on Commissioners than technical studies on the same topics. For example, during the closed-door meetings of the Commission that followed regional consultations, Commissioners would frequently invoke the previous day's stakeholder evidence in their discussions.[24]

For many observers, balance was important because regional consultations were their only direct experience with the Commission's work. They not only obtained their primary impression of the Commission's independence from the two-day meeting, but it was possible they would learn more from the meeting than from the Commission's final report. For example, interviews with government officials at the Egypt and Vietnam meetings revealed that they attended the two-day meeting to gather information about state-of-the-art dam-building or alternatives. Few were conscious of, or looking forward to, the WCD's final guidelines. Moreover, language interpretation at the meetings made the content accessible to a local audience, whereas ongoing news of the Commission's activities posted in English on the WCD's website was less accessible.

Participation versus Consultation

The WCD followed emergent international norms by undertaking a consultative process. At the same time, the Commission determined that the contentious nature of the topic and the history of struggle required strict rules of participation. Far from allowing free-flowing public dialogue at its consultations, the Commission carefully handpicked presenters for the regional consultations and timed the speeches strictly. The result was a consultation that was structured around a series of testimonies, lending the Commissioners the air of judges who would weigh the evidence in an independent manner.

This somewhat rigid approach did vary according to who held the chair. For example, in all cases only Commissioners were permitted to question the speakers. However, at some consultations, observers were allowed to comment from the floor. Personal access to Commissioners during lunches and coffee breaks varied widely from meeting to meeting. Indeed, as one Commissioner noted, the style of the hearings changed in a kind of "action-reaction" process of learning.[25] After the enthusiastic participation of hundreds of displaced people took the Commission by surprise in São Paulo and drew criticism from members of the dam establishment, the following regional consultation in Cairo was managed on a much tighter rein. Observers in Cairo had no chance to comment in the public proceeding but were, instead, urged to provide submissions in writing.[26] In retrospect, the Commission could have done a better job of clarifying the scope for participation before the meetings, in order to manage expectations.

Consultations were structured as opportunities for public testimonies, rather than for stakeholder dialogue.

The highly structured meetings surprised many participants who had hoped for a more open exchange of views on large dams. Early WCD publicity materials that called for a consensus-building approach raised expectations of a free-flowing exchange at meetings and lively dialogue with Commissioners.[27] The assessment team found this expectation to be particularly strong at the São Paulo meeting.[28] The emphasis on testimony rather than debate was a potent example of how the process was weighted more toward Commissioner findings and less toward promoting a broad debate among stakeholders.

By contrast, the country meetings, which were designed to provide focussed feedback to the WCD on thematic and case study papers, had less the flavour of hearings. These events provided greater opportunity for spontaneous debate than did the tightly organised regional consultations, but their geographic scope was also narrower.

Overall, the WCD's decision to manage the large, regional hearings tightly was a reasonable decision because it prevented partisan voices from dominating. However, the consultative process also highlights the trade-offs faced by the organisers of these and future events. There is not necessarily one correct way of facilitating public meetings for diverse participation. On the one hand, facilitating direct dialogue between broader stakeholder groups offers the possibility of advancing mutual understanding; the WCD accomplished this only in a modest sense. On the other hand, structuring stakeholder meetings so that they deny more substantive debate among participants helps keep the events politically balanced; this was the WCD's choice.

The Public Face of the WCD: The Media Strategy

The Commission's media strategy focussed on gaining coverage for meetings after they occurred and for the overall work of the Commission. The Commission may have missed an opportunity by failing to integrate its media strategy with its strategy for mobilising submissions and attendance at the consultations.

The primary tools deployed by the WCD to foster inclusion were its newsletters and website. The effectiveness of these vehicles for information was enhanced by arrangements with specialised publications and institutions to amplify the WCD's call for input. For example, the WCD inserted information on its process in industry publications, such as *Hydro Review Worldwide* and *International Water, Power and Dam Construction*, to reach industry groups. Similarly, various civil society groups and networks circulated information about the WCD through their own electronic networks. As with efforts to broaden participation at the regional consultations, the willingness of outside actors to play a constructive role was of considerable benefit to the WCD process.

Local media was not a central element of the WCD's outreach strategy, particularly in its early days. Although media packets were put together at the time of regional consultations, these were assembled by local consultants who adapted and translated WCD information, without much WCD control over the final outcome.[29]

By the time of the East and Southeast Asia consultation in Hanoi in February 2000, the last regional meeting, the WCD had refined its outreach model somewhat. In this case, the Secretariat hired a member of IUCN-Hanoi to work full-time on Vietnamese language outreach for the six months leading up to the event. This model solicited substantial input from a cross-section of Vietnamese stakeholders (within the framework of a small NGO sector), and generated some media announcements about the meeting before it happened.[30] This model of an integrated outreach strategy that incorporates both presswork and a call for submissions to general stakeholders would have applied well to the entire WCD process and provides a lesson for future processes.

Sustaining high profile media attention proved to be an uphill battle.

For the most part, the central goal of the media strategy was to achieve a high profile with national and international media. This was particularly true in the early days when media attention was perceived as a useful way to attract financing for the WCD's process. However, because of the absence of consistently interesting newsworthy events during the life of the WCD, sustaining high profile media attention proved to be an uphill battle. Other than the launch event itself, the most newsworthy stories during the WCD's tenure proved to be controversies surrounding the regional consultations, in particular the cancellation of the Indian consultation.

The media strategy was only partly successful at shaping the Commission's public image. The media's filters significantly influenced its portrayals of the WCD. The Secretariat tried to focus media attention on the uniqueness of the process itself, but this theme did not get off the ground.[31] Press releases prominently noted the Commission's adherence to principles of good governance,[32] but the media did not pick up on the importance of good process to good outcomes. Similarly, only 3 percent of WCD press releases focussed on negative stories about dam-related displacement and environmental problems, but a substantial 32 percent of media reports sampled for this study focussed on such negative stories.

The voices of stakeholder groups were also important in shaping media coverage and, by extension, the WCD's public image. Twenty-nine percent of quotes in the media were attributed to Commissioners or Secretariat staff, but NGOs also accounted for a sizeable 21 percent of WCD-related quotes—more than either governments (12 percent) or the private sector (6 percent). These figures suggest that NGOs were more effective in getting their messages heard in the media than were other stakeholders.

During the latter half of the WCD's tenure, attempts to attract media attention focussed on sustaining interest in the WCD as a source of credible information on dams. Specifically, the WCD sought to raise media and public interest by putting out issue-specific and geographically-focussed press releases. However, this strategy ran the risk of alienating stakeholder groups who felt that the WCD was prematurely declaring conclusions and violating its self-imposed stricture against arbitrating ongoing controversies. In the words of a Secretariat member, "The distinction between [the WCD's] input and output was blurred."[33] For example, some groups interpreted a press release on the Pak Mun Dam (Thailand) as a judgement by the WCD on the dam, while the release was intended to raise the relevance of the WCD for Thailand. In another example, industry groups questioned the credibility of a press release that estimated the number of people displaced by dams and deemed the release premature.[34]

The media strategy was only partly successful in shaping the WCD's image.

These examples suggest a more general problem faced by the WCD in attracting and sustaining media attention—media attention and maintenance of stakeholder trust in the process were often at odds. To attract media attention, the WCD had to release newsworthy information, which, for example, shed new light on a heated debate or addressed an ongoing controversy. However, the sharper and more newsworthy the release, the more likely it was to alienate stakeholders, and thereby undermine attempts at sustaining the multi-stakeholder dialogue.

Conclusions

Public consultations are an emerging norm for development decision-making, especially international multi-stakeholder processes. As with the

Lessons for building credible multi-stakeholder processes

- Hold public hearings to foster inclusion of diverse viewpoints.
- Establish early the objectives of public consultation and make sure to communicate these clearly to participants, including how their contributions will be used.
- Consider providing modest financial resources to selected local groups to reach out to communities and other relevant groups who would not otherwise hear about the process.
- Provide financial support to community representatives and other less-resourced groups to enable them to travel to meetings, so that the scope of their participation is equivalent to that of government, business, and better-resourced groups.
- Use international networks to disseminate information about events, but also exploit country and regional networks and the mass media, where possible, to reach popular audiences.
- Integrate media and general outreach strategies fully.

WCD, they can provide an important opportunity for otherwise disenfranchised voices to have their say. Moreover, in-person presentations at consultations provide a directness and immediacy to commission deliberations that cannot be substituted for by other forms of knowledge gathering. Finally, consultations can contribute greatly to the reality and perception of inclusiveness, which buttresses the legitimacy of a commission. However, consultations can also be counter-productive, if the tremendous political weight of a global commission comes to bear inappropriately on a local or national controversy and the commission is seen as unbalanced or taking sides.

To ensure that consultations are inclusive and balanced, commissions will have to shoulder considerable logistical costs. As was the case with the WCD, the use of organisational nodes to disseminate information and mobilise participation can help ease this burden. However, in the future, a conscious strategy of collaboration with stakeholder groups may require that a commission build in more funds to assist stakeholders—particularly under-resourced NGOs—in outreach. Such a strategy also has its limits and can at best be a complement to, and not substitute for, a secretariat's outreach.

Much of the credibility fostered by public consultation derives from providing equal opportunity to all major stakeholder groups to participate. The WCD did this well: by pre-selecting speakers for its regional meetings for political, geographical, and topical balance. The WCD levelled the playing field for participation by offering travel stipends to all speakers to attend the regional event. Such enabling mechanisms for less affluent stakeholders will be essential to the credibility of future processes. The disadvantage of the carefully constructed and controlled meetings was that they lost the Commission some credibility as a forum for exchange. Participants were not able to debate directly in the WCD consultations; this confused those who thought the objective of the process was a broader consensus-building. As the meetings progressed, the Commission became increasingly explicit about the scope of participation it allowed. Such clarity about the way people can participate, and how their contributions will be used, is essential to the legitimacy of any multi-stakeholder or broader policy process.

The WCD may have raised greater popular awareness of its work and increased meeting attendance had it tied its media strategy more explicitly to the publicity for consultations. It accomplished this only late in its process. This experience suggests the importance of multi-pronged outreach strategies that use country- and region-specific networks and the mass media to promote diverse participation.

Endnotes

1. *Dams: Official Newsletter of the World Commission on Dams*, No. 6, April 2000.
2. Interview with senior Secretariat staff, December 1999.
3. The Commission was forced to cut back on some of its original plans in developing regions; for instance, the Commission had to drop the idea of a South African regional hearing. In that particular case, NGOs arranged a hearing with private foundation funding as a direct input to the Commission's work instead. The history is documented in the Environmental Monitoring Group's record of the proceedings.
4. Based on a comprehensive survey of Indian press articles on the WCD, 1997-2000, by Ramananda Wangkheirakpam.
5. Virendra Kumar reporting for the *Indian Express*, Ahmedabad, 9 September 1998.
6. Special correspondent reporting in *The Hindu*, 5 September 1998.
7. Interview with Commissioner, 27 February 2000.
8. Interview with senior Secretariat staff, December 1999.
9. Commissioners' contextual remarks to the meeting participants in Hanoi, February 2000. Vietnam is expected to install a further 11,000MW in hydropower capacity in the period 2001-2020.
10. Interview with Commissioner, 19 November 2000.
11. Interview with staff of Egyptian human rights group, 8 December 1999.
12. For example, the Egyptian NGO presentation to the WCD at the regional consultation concerned this NGO's efforts to help the government rehabilitate irrigation infrastructure. Also, interview with independent Egyptian academic and environmental consultant, 10 December 1999; interview with director of Egyptian environmental NGO, 11 December 1999.
13. One local government official, whose participation was supported by the NGO, was also present.
14. Interview with presenter at the regional consultation, February 2000.
15. Interview with official from the Egyptian Department of Irrigation, 8 December 1999.
16. WCD Secretariat, "Call for Submissions," announcements to WCD Forum members, July 1999-January 2000.
17. Interview with senior Secretariat staff, February 2000.
18. Information drawn from Flávia Braga Vieira, "Brazil's Dam-Affected People's Movement and the World Commission on Dams." Background paper prepared for the WCD Assessment, January 2001.
19. Interview with NGO activist, MAB's IV National Congress, November 1999.
20. In Cairo, however, the meeting took place at a deluxe hotel. According to NGO participants who sought more modest alternatives, the hotel was not located near other, lower priced accommodation. International NGO observers were able to muster the funds to share rooms and attend the proceedings, but the cost of rooms remained contentious. The second stakeholder meeting for the Tarbela case study in Pakistan was held in a five-star hotel, at which local NGO representatives refused to meet. Similar constraints forced community or NGO stakeholders to turn away from the Oregon (U.S.) and Oslo (Norway) meetings.
21. Exceptions were the Commissioners from international business and government backgrounds, who attended relatively fewer consultations. As a result, their sectors may have appeared partly under-represented to observers.
22. Interview with Commissioner, 20 January 2000.
23. International Rivers Network wrote to the Secretary-General to protest the exclusion of Dai Qing, the Three Gorges Dam's most outspoken critic, from the agenda. The only Chinese presenter scheduled to speak came from the Institute of Agricultural Economics in China, and his paper was solidly pro-government and pro-dam. The other presenter on Chinese dams was the WCD's own foreign consultant, who had written a stock overview based on government statistics.
24. Interview with senior Secretariat staff, 6 November 2000.
25. Interview with Commissioner, December 1999.
26. Attendance at the Cairo consultation, December 1999.
27. Manuel Pulgar-Vidal, background paper to the WCD assessment on the Latin America regional consultation, August 1999; Braga Vieira, 2001.
28. Braga Vieira, 2001; Pulgar, 1999.
29. The result was some tragi-comic lapses. For example, the early publicity materials for the WCD in Egyptian referred to the World Commission on Cisterns, an error that was fortunately captured and corrected before the Cairo consultation.
30. Such as an announcement on Vietnamese state television.
31. Interview with WCD media staff, 7 April 2000.
32. Independence, openness and transparency, knowledge-driven, accessible, representing a spectrum of views, and informed on issues and alternatives. All six of these principles were highlighted in 60 percent of WCD press releases.
33. Interview with senior Secretariat staff, 6 November 2000.
34. Interview with industry representative, 17 November 2000.

Chapter 7

Commission Dynamics: Narrow versus Broad Consensus

Two key features of the WCD that we explore in the previous chapters—breadth of representation on the Commission and a work programme based upon good governance principles—enabled the Commission to produce its report. The Commission's diversity enhanced the legitimacy of its deliberations, even as it made producing a consensus report particularly challenging. At the same time, an open and inclusive work programme was necessary to build and maintain trust among Commissioners and to allow each Commissioner and the Commission as a whole to maintain connections with stakeholders.

This chapter examines the dynamics within the Commission that led to consensus, and the degree to which these interactions can be planned for and structured into future multi-stakeholder processes. This chapter also foreshadows international reactions to the final report, which are discussed in Chapter 8. The broad representation on the Commission and its attempts to achieve good process were intended to build constituencies for adoption and implementation of its final report in the greater stakeholder community. However, the Commission's choices about where to seek consensus led it toward consensus-finding among its own members, rather than consensus-building or even the evolution of shared understandings among Forum members and larger networks. We look at the choices that led the WCD in this direction, as a precursor to the discussion about adoption and compliance in Chapter 8.

Toward Commissioner Consensus

For the Commission to be successful, the diverse Commissioners had to develop a shared interest in achieving consensus. This was not a small task. Commissioners brought very different perspectives and histories, and in some cases considerable wariness of the process and of each other. In this context, Commissioners have emphasised the central role of a balanced and strong Chairperson in steering Commission deliberations.

Commissioners put aside difficult issues for discussion later.

As a government minister and an academic with a history of political struggle, Chairperson Kader Asmal brought considerable authority to his stewardship of the WCD. In chairing the Commission, he drew on the idea of "sufficient consensus," an approach used by the African National Congress during anti-apartheid days to bind coalitions in the party and to build relations with other parties. He encouraged Commission members to focus first on issues around which they could forge rough consensus and to table difficult issues for later discussion.[1] The exercise of veto by one member over the entire process was not an option. This approach was meant to decrease the likelihood of dissenting reports and cultivate a sense of individual and shared investment in the final principles. By persuading Commissioners to invest incrementally in the process of mutual learning and relationship-building, Professor Asmal intended to make it ever more difficult for a Commissioner to walk away from the process. He told Commissioners that their enemies were no longer each other, but the sceptics who said they could not finish their work.[2]

The first meeting of the Commission was reportedly fraught with tension. Each member of the Commission introduced him or herself to the others, laying out his or her professional and personal experiences. A Commissioner who worked with displaced people showed pictures of terrible resettlement conditions to the Commissioner whose corporation was supplying engineering equipment to that dam.[3] Although this beginning did not augur well for conciliation, one Commissioner noted that it was necessary to "get this out of the way."[4] Once these introductions were made, the Commissioners could shift their focus to the mandate and the work ahead. Over time, there developed within the Commission what Commissioners describe as a remarkable climate of mutual respect on a personal level,[5] and a growing willingness to listen to the others.[6]

Dr. Jan Veltrop, a former President of the International Commission on Large Dams (ICOLD), was commended by all of his colleagues for wedding a lifetime of experience in dam building with an extraordinary willingness to listen to different viewpoints and seek out new information. He was also tireless in his work during the life of the Commission and attended almost every field visit and consultation. Dr. Veltrop was a replacement for another former president of ICOLD who had originally been appointed to the Commission.[7] Whether the original Commissioner from this key constituency would have played an equally positive role can only be a matter of speculation. However, at least some Commissioners suggested that had he stayed on, the Commission as a whole would not have progressed toward the final framework. These observations highlight the importance of individuals and, to some extent, serendipity in the success of the Commission.[8]

At the other end of the political spectrum, Medha Patkar, the social movement activist from India, is credited with retaining her convictions while maintaining enough flexibility to allow the Commission to move forward. Some of the Commission's early difficulties with the India consultation have been attributed to Ms. Patkar's pushing her interests too strongly.[9] However, her subsequent willingness to negotiate in the framework of the overall Commission's goals enabled the body to accomplish more than many thought possible.[10] Ms. Patkar did append a "comment" to complement her signature on the final report, in which she argued that the report's scope was inadequate and did not address the broader structural problems with associated development, of which dams were but one symptom. However, her "comment" was couched in the context of endorsing the process and its outcomes.

The most significant change in the Commission itself occurred halfway through the process when Commissioner Shen Guoyi, of China's Ministry of Water Resources, abruptly resigned.[11] Ms. Shen's departure certainly held implications for the Commission's external image, for she was at once a senior government official, a Southern voice, and the only woman who was not from an NGO or community background. Her presence on the Commission had also signified co-operation from the world's most prolific dam-building nation. Although the WCD Chairperson requested a replacement Commissioner, the Chinese government did not respond. Ms. Shen's absence would later amplify criticisms that the Commission received after the launch of its report—that it was biased toward the North and that it had failed to consult Southern governments adequately.

Although one can only guess the effect of Ms. Shen's departure on Commission negotiations (after all, the Commission had not even begun to prepare the final report when she left), it is likely that her resignation influenced the Commissioners' ability to reach consensus—as well as the nature of the final report. While the Chinese government did not make any formal statement of disagreement regarding the substance of the Commission's work at the time of Ms. Shen's departure, a later memo from the Chinese government to the World Bank indicated that it disagreed with the positions of most Commission members.[12] *(See Chapter 5, Box 5.4.)*

Toward Consensus: The Process of Shared Learning

The Commission's success in completing a consensus report was largely a result of a two-year shared learning process. Had the Commissioners sought to establish common ground early, their debates would have been based on prior experience rather than shared evidence. Hence, they decided to cast the net wide and avoid discussion of development effectiveness criteria up front. A focus on gathering different views enabled the Commissioners, quite simply, to get on with each other.[13]

"I think the decision not to dive into the work programme was a smart one," said one Commissioner. "A joint fact-finding, with a common knowledge base...allowed us to build trust. It's been very collegial, congenial."[14] The Commission's focus on listening and learning was intended to bring the members toward a consensus of the highest common denominator.[15]

The process of shared learning made Commissioners more determined to complete their mandate.

As we documented in Chapter 5, the process of negotiating the Commission's work programme democratised, or opened up, knowledge creation to go beyond scientific sources and provide a space for grassroots voices. This would not have been possible without the presence of community and indigenous peoples' activists on the Commission itself. Meanwhile, the presence of practitioner Commissioners with extensive fieldwork experience also broadened the whole Commission's view. Joji Cariño is often cited as having made a notable contribution to the Commission's collective accomplishment in addressing indigenous and tribal peoples' issues. Deborah Moore's background in environmental flows, Thayer Scudder's in resettlement issues, and José Goldemberg's in the energy sector are cited as having compelled and advanced collective thinking.[16] Commissioners prepared special briefing papers on their areas of expertise and presented these at Commission meetings. They took special responsibility for commenting, on an ongoing basis, on parts of the work programme in which they had a particular interest. Later in the process, they formed subcommittees to consider focal issues in depth.

The unfolding work programme provided the opportunity for Commissioners to absorb and debate new material together, via common readings, public consultations, and shared experiences with engineers, activists, communities, and decision-makers. At times, Commissioners had first-hand experiences that broadened their perspectives. For instance, the engineer on the Commission who had not been exposed to the conditions of resettled communities was reportedly much moved by observing and hearing from these affected communities.[17] This gradual process of learning, organised around grappling with the same set of material and processing the same experiences, equipped the Commissioners to confront difficult decisions in the last few months of their work.

"Why did I sign off?" said one Commissioner, on the eve of the report's launch. "Because of an early desire on my part, and I know on the part of others as well, to succeed. Later, among all of us there was a determination to complete our mandate. Surely the final report represents some measure of compromise for many of the Commissioners, if indeed not for all of them. We realised what would be at stake if no agreement would be reached."[18]

Commissioner Consensus versus Stakeholder Consensus

A "consensus driven approach" was one of the WCD's guiding principles and indeed, part of the Gland mandate. But the Commissioners interpreted consensus-building as focussing on interactions with each other. Building consensus in the broader stakeholder community was a longer-term objective for which the Commission's work would lay the foundation. As principal advisor Anthony Dorcey had observed in the early days of the Commission's formation:

> "It is proposed in the Gland report that from inception to conclusion the full diversity of stakeholders should be involved in a process that is transparent and designed to reach consensus to the greatest extent possible. While this reflects a growing trend in new collaborative models of governance for sustainability, putting those principles into practice raises immense challenges for the commission. Practical considerations of time and resources will generally make it impossible to go much beyond achieving goals of transparency to build consensus in its multi-stakeholder consultation processes.
>
> *Recommendation:* The terms of reference document should make clear that while the immediate goal of the commission focuses on building consensus among the commissioners on conclusions and recommendations, the longer term goal is to provide a foundation for building consensus in large dam decision-making processes."[19]

The Commission's work programme reflected this emphasis. For instance, as we described in Chapter 6, the public consultations were designed and presented as information-gathering opportunities for the Commission, not occasions for exchange, dialogue, or consensus-building among stakeholders in the broader community.

The Commissioners focussed on building consensus among themselves, not among stakeholders at large.

Because there were few opportunities for direct exchange among a broader number of interest groups under WCD auspices (Forum meetings provided one such opportunity), many interest group efforts were directed toward influencing the Commissioners themselves or lobbying members of the Secretariat who were involved in the day-to-day management of the work programme. Thus, for those stakeholders who were actively engaged in monitoring and providing input to the work programme, the Secretariat and Commission acted as mediators. Because it was involved in the day-to-day implementation of the work programme, the Secretariat felt much more of this pressure than the Commission did.

Constituency Ties, Constituency Drift[20]

As explored in Chapter 4, the Commission's legitimacy derived from the range of networks, views, and positions collectively represented on the Commission. Individually, for many Commissioners, credibility rested on continued ties with and support from networks. At the same time, each was empowered to act in his or her personal capacity, to negotiate freely toward consensus. This simultaneous need for connection and freedom established a tension for many Commissioners.

In some cases, the ties to constituents were particularly strong and were reinforced throughout the life of the Commission. This was especially evident in the case of Commissioners from non-governmental backgrounds, who consulted with international NGOs and dam-affected peoples' representatives before each Commission meeting. These informal consultations allowed them "to keep in touch," and ensure that they didn't "sign off on things people aren't comfortable with," in the words of one Commissioner.[21] "I have a responsibility to represent the point of view of a particular constituency," said another, who noted that peoples' movements had "expressed faith" in her nomination and provided much support.[22]

Other Commissioners maintained loose networks of associates and advisors whom they used as sounding boards for issues raised in the Commission. Jan Veltrop, the past President of ICOLD, turned to colleagues in the ICOLD network. Göran Lindahl, the CEO of a large private company, formed a loose network with the industry representatives on the WCD Forum. This industry group held a meeting to analyse the WCD's thematic papers in-depth prior to channelling input to Lindahl. However, this effort at co-ordination developed late in the process, long after NGOs and social movements had established well functioning mechanisms of co-ordination and feedback. This difference in mobilisation indicated the private sector's general slowness to engage with the WCD.

Commissioners had to maintain ties with their networks while maintaining enough flexibility to achieve consensus with other Commissioners.

Some members of the Commission, perhaps because of their academic or multi-sectoral backgrounds, did not consider themselves to have a defined or mobilised constituency. At least one such Commissioner constituted a "kitchen cabinet" of advisors, "all of whom have a totally different viewpoint" to act as a sounding board for WCD issues.[23]

Despite various efforts at communication and co-ordination between Commissioners and constituencies, the two-year process of shared learning and moving toward consensus inevitably required some flexibility in Commissioners' thinking. If the Commission were to produce a consensus final report at all, they had to give some ground. This raised the danger of Commissioners drifting away from their interest group networks.

For example, once the Commission had released its report, a dedicated anti-dam activist opined that "some of the WCD Commissioners were not as strong ['anti-dam'] as we thought them to be."[24] A World Bank official expressed his disappointment that a Commissioner from government agreed on a report that, in his opinion, was so far-reaching in its call for reform that it circumscribed the role of governments.[25] In the case of the representative from the private sector, Göran Lindahl, this drift was particularly strong, but largely because of his company's withdrawal from the dams business, rather than because of shifts in his position to accommodate consensus. *(See Box 7.1.)*

Commissioners' Varying Involvement

Having a commission composed of active and engaged individuals, such as the WCD, presents a host of benefits, although some commissioners face the pressures of competing responsibilities. In the WCD's case, Commissioners differed in the amount of time they were able to dedicate to Commission work. Where Commissions of the past had been stacked with eminent retirees, this Commission had only one retired person: Dr. Jan Veltrop. Although retired, he maintained an active interest in ICOLD affairs and attended almost all ICOLD meetings.[26] Some Commissioners from NGOs and community-based organisations were able to scale back other commitments so that they could devote more time to the WCD.[27] Others, such as Göran Lindahl, the CEO of a multinational corporation; Donald Blackmore, the CEO of a major river basin authority; and Kader Asmal, a sitting government minister, were limited in how much time they could make for WCD affairs.

According to Secretariat members, the varying levels of engagement of the Commissioners did not materially affect the outcome of the final report. The Secretariat members who drafted the text, as well as the other Commissioners, were highly attuned to each person's bottom line. "It's not necessarily the length of time or number of Commissioners that matters," said one senior advisor. "As long as you have the range of views there, you'll get essentially the same product. You can't move forward with something, if one major group doesn't agree."[28]

Box 7.1

Industry representation on the Commission: the case of ABB

Asea Brown Boveri (ABB), a Swiss-based international engineering company, had been closely involved in the WCD process from the days of the Gland meeting. When Commissioners were selected, the CEO of ABB, Göran Lindahl, was given a slot.

Over the life of the Commission, ABB's stake in hydropower diminished considerably. For example, ABB was a major contractor for Malaysia's controversial Bakun Dam. For a variety of reasons, including mounting NGO pressure, the company withdrew from the project.[a] Following this episode, ABB announced it would shift its focus away from traditional large-scale power plants and increase investments in alternative energy, such as wind power.[b]

By late 1999, much of ABB's stake in hydropower was gone. Other industry representatives on the WCD Forum began to feel that Mr. Lindahl was no longer the best spokesman for their interests. Mr. Lindahl offered to resign in favour of another industry representative, such as a senior representative of the Canadian company Hydro-Quebec.[c] However, given the late stage of the Commission, the Chairperson and the other Commissioners were unwilling to consider a replacement, keeping in mind the dynamic of interpersonal relations and shared learning that had evolved among the Commissioners by then.

Mr. Lindahl remained on the Commission, but with limited involvement. Indeed, the head of ABB's environment division took his place at the last Commission meeting, to sign off on the final report. In an indication of ABB's detachment from the process, no representative of the company attended the final Forum meeting, where stakeholders shared plans for charting a course forward.

[a] See "High Risk, Low Return," an NGO report on ABB's Hydropower Strategy by Nicholas Hildyard, February 1998.

[b] ABB website, www.abb.com; also Network for Alternative Technology and Technology Assessment, Open University, Renew newsletter, September-October 2000.

[c] Interview with hydropower industry representatives on the Forum, 17 November 2000.

The WCD Forum: Promise Unfulfilled?

The role of the WCD Forum was to act as a sounding board for the Commission's work and to perform an outreach function for the final report. The Commissioners intended to use the Forum to ensure that their discussions did not drift too far from the concerns and positions of stakeholders and to sustain the interest and engagement of stakeholders throughout the process. In reality, realising the full potential of the Forum proved to be a challenge, one that often foundered against the practicalities of meaningful engagement with large numbers of stakeholders in a time-bound process.

Forum Members' Varying Involvement

Forum members' involvement in the work programme depended much upon their own initiative and perseverance. Some Forum members were deeply involved in developing the WCD's knowledge base. They generated comments on terms of reference and drafts of papers, served on formal review panels, and disseminated and promoted information on the WCD to their professional networks and constituents.[29] Others were far less engaged.

A network of NGOs and social movement groups acted as constant critics and monitors of the WCD work programme. Later in the process, a dams industry group forged a similar network, which, over time, substantially increased its scrutiny of and its voice in the process. These three broad categories of stakeholders tended to be motivated by strategic concerns. Social movements and NGOs were partly driven by a concern to prevent industry from dominating the agenda. Some members of the dams establishment openly admitted that their motivation for involvement was to "limit the damage" by the WCD to industry interests.[30]

By contrast, other Forum members, such as the Inter-American Development Bank, did not attend any Forum meetings, but were active behind the scenes and between meetings in arranging consultant input to the WCD's work programme.[31] Many of the policy research and academic institutions on the Forum largely limited their engagement to participation in Forum meetings. As a general observation, those Forum members who had a direct stake in the outcome of the Commission—public and private industry engaged in dam building and operation, and social movements of affected communities and their allies—tended to put the most effort into the process.

Forum members' inputs into the work programme occupied countless hours of Secretariat time. Secretariat staff members were required to respond to requests while remaining completely even-handed toward the diverse groups on the Forum. Initially, the Secretariat arranged a system by which staff members were tasked with contacting Forum members on a regular basis. This system did not work, because the Secretariat was too busy "to engage them on issues they didn't bring up."[32] Instead, the Secretariat designed review processes for thematic and case study papers for multi-stakeholder input and responded to specific Forum requests as they came up.

Communicating with the Forum occupied countless hours of Secretariat time.

Secretariat staff found active Forum members to be helpful, because such members knew the work programme well. They provided "very targeted" comments to the Secretariat's work, thus acting as an effective sounding board.[33] At the same time, the staff's role as mediators between the various interest groups was a high-pressure vocation that gave staff "the feeling of living in a fishbowl."[34] As they recall, it was hard for outsiders to appreciate the degree of consultation that took place within the Secretariat and between the Secretariat and key opinion-shapers on the Forum before decisions were made about activities that would influence public perceptions.[35]

"All they see is a letter of invitation, a meeting agenda and assume it's top-down. They never see the discussions we have about who should be invited, and in what order. To get to a letter of invitation and an agenda is often the result of five or six meetings, informal phone conversations with key people… In the back of our minds there's always: What will the Commissioners think? What will the stakeholders think? We're very aware of the boundaries."[36] The intense pressure upon the Secretariat holds lessons for future multi-stakeholder processes that seek to gather new data on contentious issues. The WCD's experience suggests that the personal qualities of Secretariat members, such as diplomacy, tact, and fairness, are critical to the legitimacy of such processes.

The Commission's Use of the Forum

The composition of the WCD Forum well illustrated the Commission's efforts at inclusiveness. The Forum included organisations that had been engaged in pitch battles over the preceding years about specific dam projects. Dam financiers, builders, opponents, and affected communities all had a seat at the Forum's table. Bringing these groups together over a sustained two and a half year period was a notable accomplishment.

Yet, although this gathering was unique, the scope for stakeholder dialogue during structured Forum meetings was limited. The Forum only met three times, including one meeting after the release of the WCD report. Of the three meetings, only the second provided an opportunity for Forum members to review the expanding knowledge base together and use these discussions to further collective understanding. As described earlier, the focus of the Commission's design was to promote consensus in the Commission itself, as the first hurdle for a legitimate report. To serve this overarching objective, practical considerations of Secretariat time and resources required that Forum meetings be limited. Although this choice is understandable, the opportunity for more direct dialogue within the Forum was lost.

Forum members complained that they were not given sufficient opportunity to comment on the Commission's work.

Commissioners and Secretariat staff both praised Forum meetings for providing valuable reality checks to the Commission's evolving views on large dams and alternatives.[37] Commission members emphasised the Forum's helpfulness in providing feedback on the Commission's ideas as it prepared to debate the final report. However, the utility of the Forum was not as evident to Forum members as it was to the Commissioners and Secretariat. Both in confidential interviews and in open meetings after the launch of the WCD report, Forum members complained that the opportunity for feedback presented by Forum meetings was largely wasted. They criticised the Commission for not producing and sharing a set of interim synthesis findings with the Forum, which could have served as a concrete basis for soliciting their feedback. Given the WCD's focus on a Commissioner consensus, rather than a stakeholder consensus, sharing interim findings raised several thorny issues for the process. We turn to these issues next.

From Workplan to Final Report

Transparency in Preparation of the Commission's Report

In pulling together the many and disparate strands of the knowledge base, the Secretariat was under enormous time pressure. The Commission's original plan was for the Secretariat to produce synthesis outputs from the knowledge base corresponding to the three objectives of the WCD: a review of development effectiveness; a framework for options assessment; and a framework for criteria and guidelines around dams. The intent was that the Commission would use these to formulate its final assessment of dams' performance and forward-looking guidelines.

In the end, it took longer than the Secretariat and Commission expected to finalise the knowledge base studies. In fact, some commissioned papers were not available in final form for use in the Commission's deliberations. Moreover, although available on request, relevant submissions from the public were still being appended to thematic papers as the Commission's final report was released, suggesting that they were not readily accessible to the Commissioners. As a result of these time delays, synthesis outputs were not circulated to the Forum. Instead, much of the material that would have gone into these outputs was wrapped into the final report.

There was much confusion as to whether the Commission's failure to circulate synthesis outputs represented a breach of faith. Some industry stakeholders and World Bank staff, in interviews conducted after the report was released, said they had expected that the synthesis product would be open to public review and comment.[38] The Secretariat noted that their work programme never stated that the outputs would be circulated, merely that they would be produced. However, given the Commission's commitment to transparency and the willingness to circulate drafts of commissioned papers, it was reasonable for Forum members to expect that synthesis outputs would be shared with them. The trade-offs involved in additional consultation are reflected in WCD participants' comments in Box 7.2.

The value of producing and disseminating an interim report would have been to show stakeholders, particularly Forum members with their ability to influence public opinion, the direction of the Commission's thinking, and would have presented an opportunity to make constructive criticisms to increase the acceptability of the final product. For example, industry stakeholders insist that review of synthesis outputs would have enhanced the technical quality of the final report.

An interim report would have revealed the direction of the Commission's thinking.

Ultimately, the Commission's choice not to devote time to preparing and soliciting feedback on synthesis outputs was a choice to prioritise Commissioner consensus over stakeholder consensus. To defend the integrity of the Commissioners' deliberations, the Commissioners chose to pay the price of reduced Forum engagement because several members of the Commission and Secretariat feared sharing even interim products with the Forum would stall the Commissioners' negotiation process.[39] This option carried a price. The decision diminished the Forum within the process, cost the Commission a measure of goodwill and credibility, and lost some potential ambassadors for its message.

Some Forum members raised similar arguments about preparation of the final report, as distinct from the interim synthesis outputs. Indeed, there was intense secrecy around the Commission's final report once it was completed, with no opportunity for review outside of Secretariat staff and Commission members. Faced with this lack of access, some Forum members charged that in the interests of transparency the Commissioners should have shared a draft of the report with Forum members for comment prior to finalising it.

The case for opening the final report to broader stakeholder negotiation is weak. In a multi-stakeholder advisory commission, such as the WCD, it is entirely appropriate for a commission to close the doors at some point. The Commissioners had to reach and formulate their own final conclusions, based on their best judgement and on the trust placed in them during the process. Had they held the final report open to review and comment, it is likely that they would never have completed it.

Indeed, the pressure on Commissioners and the Secretariat was considerable, even without considering additional last minute views from Forum members. Given the volume of material, few Commissioners were able to read and digest the entire compilation. Instead, late in the process, Commissioners signed up for issues on which they had to concentrate to draw implications and recommendations. In the final meeting of the Commission, members "formed groups to go over, revise, then reject whole chapters."[40]

Box 7.2

Participation and transparency versus time

"If you open up the planning too much so everyone has a say in every point of the process, it's all going to bog down. It's a better idea to get everyone to agree on who should undertake the process, and make sure it's transparent, etc...."

—a Commissioner, commenting on whether the WCD should have put out an interim report for public review[a]

"I'm a big fan of time-limited processes, they force you to grapple. They are massive challenges to processes that want to be participatory. Participation is where it falls out."

—a Commissioner, commenting on the overall time frame of the WCD process[b]

"They were supposed to bring together three synthesis reports. We never saw some of the most important knowledge base products for peer review. I believe the Secretariat could have had a peer review or hearings. Couldn't a synthesis step have helped a bit? It would've been a shorter, much more concise report. Two years was a break-your-neck kind of exercise."

—a Forum member, commenting on the final synthesis and report-writing process[c]

[a] Interview with Commissioner, 19 March 2001.

[b] Interview with Commissioner, 10 April 2000.

[c] Interview with Forum member, 22 January 2001.

Commissioners also relied heavily on the Secretariat for support and preparation of drafts. In contrast to former Commission reports that have been authored primarily by a Secretary-General, many people wrote this report. "Drafts went around about 30 times, marked up by everyone," noted one senior advisor.[41] According to staff, they re-created debates among themselves that mirrored the debates on the Commission itself.[42] Because Secretariat members were together in one location and were responsible for day-to-day handling of material, they had discussed more material together than the Commissioners had. The Secretariat did much of the writing, and it was helpful that Secretariat members represented sectoral diversity when they took on this task. This practical contribution toward forging commissioner consensus reinforces the importance of a diverse Secretariat in building a commission's legitimacy in the eyes of stakeholders. *(See Chapter 4.)*

Cultivation and Undermining of Stakeholder Support

Once the report was completed, Secretariat and Commissioners stepped up their outreach efforts to constituents who appeared particularly sceptical of the WCD process. In the four months before the report's release, Commission members made presentations at major congresses of ICOLD,[43] the International Hydropower Association (IHA),[44] the International Commission on Irrigation and Drainage (ICID),[45] and the World Conservation Union (IUCN).[46]

Despite these efforts, stakeholder buy-in to the report was compromised during the lead-up to the launch by leaks of the final report. It is standard practice for agencies to release official documents to the media ahead of the official launch date. In this case, some members of the Forum, particularly those from industry, were distressed to receive calls from the press requesting comment before they had seen the report themselves.[47] This frustration was enhanced by the tenor of the media's questions, which focussed on a perceived negative tone in the report with regard to large dams. By contrast, the NGO advocacy community was well prepared for the launch and had managed to obtain a copy of the report through a leak, much to the displeasure of industry groups. Industry groups grumbled publicly about what they perceived as a breach of faith, although some also privately said they had simply been out-flanked by the more media-savvy NGOs.[48]

Conclusions

Lessons for building credible multi-stakeholder processes

- Build a process around shared learning and relationship building among commissioners to enable them to transcend initial differences in perspective and experience.
- Use advisory forums to create structured opportunities for multi-stakeholder input to the process. If forum members are to be used as ambassadors, they must be briefed regularly on the substance of the developing product to gain their support.
- Respect stakeholder expectations to comment on work products, given current norms of transparency. Weigh stakeholder expectations for comment against the risks of disrupting a fragile consensus.

Organising the work of a commission around a process of shared learning helps build the basis for commissioner consensus. Shared learning allows individuals to develop personal relationships and invest collectively in a process, thereby raising the personal and professional costs of withdrawal. Skilful management within the Commission along these lines was an important contribution to the WCD process. At the same time, this approach by no means guarantees that all commissioners will stay with the process. The withdrawal of a WCD Commissioner from China, under circumstances that remain ambiguous, illustrates the political trade-offs that global commissions must manage.

Efforts toward consensus can place considerable strain on commissioners' time and on their ties with their constituents. To invest sufficiently in the shared learning process, commissioners must be able to devote time to digest materials and work with colleagues on the commission. An effective and engaged secretariat can mitigate some demands on commissioners' time. Commissioners also face the challenge of retaining their flexibility to compromise in the interests of a consensus, even while maintaining their credibility with their networks and constituents.

Establishing an advisory board, such as the WCD's Forum, is an idea with great potential. In the WCD, this potential was only partially realised. The limited opportunity for dialogue between Commissioners and the Forum and among Forum members constrained the Forum's ability to act both as a sounding board and as a mechanism for building stakeholder consensus. At the same time, given the trade-offs between investing time and resources in developing a Commissioner consensus and making incremental progress toward a consensus among all stakeholders, the WCD likely made the right choice in focussing on Commissioners.

Commissions and other policy processes face questions about when is the best time to consult the public. The Commission did not release an interim document that revealed the direction of its thinking and likely emphasis of its final report. Many Forum members saw this as a lack of transparency and openness. Significantly, this decision prevented the Commission from advancing the debate over key issues further within the Forum. This decision was to have ramifications for reception of the final report—which was more hostile than it might have been had an interim product been shared and debated in the Forum. The experience suggests that future multi-stakeholder processes, while respecting the importance of commissioners as final arbiters, should consider the value of circulating interim, not final, findings to an advisory forum for comment.[49]

Endnotes

1. Interview with advisor to Professor Asmal, April 2000.
2. Interview with Commissioner, 8 December 1999.
3. Interview with senior Secretariat staff, 8 December 1999.
4. Interview with senior Secretariat staff, 8 December 1999.
5. Interview with Commissioner, 8 December 1999.
6. E-mail interview with Commissioner, 12 November 2000.
7. The reasons given for the resignation of the original Commissioner had to do with questions of compensation.
8. Interview with Commissioner, 19 March 2001.
9. Interview with former Secretariat staff, 13 December 2000.
10. Interview with former Secretariat staff, 13 December 2000. Also, Professor Asmal singled out both Dr. Veltrop and Ms. Patkar for special praise and gratitude at the closure of the WCD process at the third Forum meeting in February 2001.
11. Ms. Shen became less engaged in Commission business in the latter half of 1999 according to an interview with a senior Secretariat member in December 1999. Her resignation was not official until early 2000.
12. World Bank, "Talking points from Government of China discussion with World Bank," internal document, 15 January 2001."
13. Interview with Commissioners, December 1999 and January 2000.
14. Interview with Commissioner, 9 December 1999.
15. Interview with Commissioner, 19 November 2000.
16. Interview with senior Secretariat staff, 6 November 2000.
17. Interview with Commissioner, 19 March 2000.
18. E-mail communication with Commissioner, 12 November 2000.
19. Anthony Dorcey, "Institutional Design and Operational Modalities for the Proposed Large Dams Commission," Stockholm Draft, 6 August 1997 (mimeo).
20. Dr. Minu Hemmati helpfully suggested the term "constituency drift" in personal communication to the authors as part of the review process.
21. Interview with Commissioner, 9 December 1999.
22. Interview with Commissioner, 27 February 2000.
23. Interview with Commissioner, 9 December 2000.
24. Interview with anti-dam activist, January 2001.
25. Interview with World Bank official, 8 March 2001.
26. Interview with Commissioner, December 1999.
27. An exception was Medha Patkar. She was often highly engaged, but at one point took eight months off WCD business to concentrate on the ongoing struggle in the Narmada Valley. This included a brief spell in jail.
28. Interview with senior Secretariat staff, 6 November 2000.
29. International Rivers Network is one such example.
30. Interview with industry representative on the Forum, 6 April 2000.
31. Interview with IADB official, 1 December 2000.
32. Interview with senior Secretariat staff, 6 November 2000.
33. Interview with IADB official, 1 December 2000.
34. Interview with senior Secretariat staff, 3 November 2000.
35. Telephone interview with senior Secretariat staff, 17 August 2000.
36. Telephone interview with senior Secretariat staff, 17 August 2000.
37. For example, interviews with Commissioners, 10 April 2000 and 19 March 2001.
38. Interview with World Bank official, 22 January 2001.
39. Interview with former Commissioner, 19 March 2001.
40. Interview with former Commissioner, 19 March 2001.
41. Interview with former Secretariat staff, 28 February 2001.
42. Interview with senior Secretariat staff, 6 November 2000.
43. Beijing, China, September 2000.
44. Berne, Switzerland, 2 October 2000.
45. Cape Town, South Africa, 24 October 2000.
46. Amman, Jordan, 4-11 October 2000.
47. Interview with Forum member, 15 November 2000.
48. Interview with Forum member, 27 February 2001.

Chapter 8

The Commission's Final Report: The International Response

The previous chapters looked at how creating a Commission based on broad representation and a process based on "good governance" principles established credibility for the WCD among a cross-section of agencies, movements, and interest groups internationally. We also looked at the trade-offs the Commission made between seeking consensus among its 12-member body and seeking to evolve a broader consensus among Forum members and networks of dam-related stakeholders. The question we address in this chapter is whether broad representation and a credible process were sufficient to ensure positive reception of the Commission's final product, a report entitled *Dams and Development: A New Framework for Decision-making*.

This chapter provides an overview of *Dams and Development*, and documents how it broke new ground in the international development discourse. We ask: What were the responses of the main stakeholder groups to the report? To what degree were stakeholders' responses based upon their perceptions of representation on the Commission, and on their perceptions of the legitimacy of the knowledge-gathering process or the consensus-building process? Based on these responses, we reflect on the implications for dissemination, adoption, and implementation of the WCD's recommendations.

Dams and Development: A New Framework for Decision-making

WCD Report Breaks New Ground

The Commission's final report, *Dams and Development*, is a consensus report of 380 pages. All 12 Commissioners[1] signed it, and Medha Patkar wrote an additional "comment." The report wraps together the required outputs of the Commission: a global review of the development effectiveness of large dams; a framework for water resources planning; and guidelines for options assessment and dam building, maintenance, and decommissioning.

The Commission's findings were much more than just a review of dams. Rather, they were judgements on the very governance and societal relations that underpin any major development project. The bulk of the Commission's Global Review of Large Dams was dedicated to analysing the performance of dams based on the questions in its case studies and survey: What were the projected versus actual benefits, costs, and impacts of large dams? To what degree had dams delivered on developers' promises or fallen short? The Commission concluded that large dams vary greatly in delivering predicted water and electricity services and related social benefits. Irrigation dams have tended to fall short of physical and economic targets. Hydropower dams "tend to perform closer to, but still below, targets for power generation, generally meet their financial targets but demonstrate variable economic performance relative to targets."[2] The history of large dams reveals a "pervasive and systematic failure" by governments and developers to assess the range of potential negative impacts and to put adequate mitigation and compensation measures in place.[3]

In analysing the causes of these failures, the Commission pinpointed inequitable power relations within and among nations and closed decision-making processes. By normative standards, the positive contribution of dams—to irrigation, domestic and industrial consumption, electricity generation, and flood control—had been "marred in many cases by significant environmental and social impacts which when viewed

Box 8.1

The role of governance in dam-related conflict

"Some may feel this Report makes water use decisions even more difficult by raising the bar higher as we do, a government must exercise more energy and creativity to reach a sustainable result. But in truth we make those decisions easier, for we show clearly which, how, where and why decisions can either work well or fail to deliver.

For that reason, I assert that we are much more than a 'Dams Commission.' We are a Commission to heal the deep and self-inflicted wounds torn open wherever and whenever far too few determine for far too many how best to develop or use water and energy resources. That is often the nature of power, and the motivation of those who question it. Most recently governments, industry and aid agencies have been challenged around the world for deciding the destiny of millions including the poor, or even popular majorities of countries they believe to be helping."

— Kader Asmal's preface to Dams and Development

from today's *values*, are unacceptable" (authors' emphasis).[4] Chairperson Kader Asmal's preface to the report further underscored the unacceptability of the decision-making processes behind much dam construction. *(See Box 8.1.)*

The Commission proposed using three United Nations instruments, on human rights and the right to clean environment and development, as a bridge between its evaluation of past mistakes and its prescription for the future. Specifically, the Commission evoked the Universal Declaration of Human Rights (1948) and its subsequent elaboration in the Declaration on the Right to Development adopted by the UN General Assembly (1986), and the Rio Principles agreed at the UN Conference on Environment and Development (1992).

The link to UN instruments was highly strategic. The Commission had been tasked with creating "internationally acceptable" guidelines for the planning, construction, and maintenance of large dams. It turned to the values set forth by formal representatives of the people—the member states of the United Nations—to ground its recommendations.

As the product of a multi-stakeholder entity with no formal legal status, the Commission's report was destined to stand as an advisory, non-legal document in the international arena. Yet, by placing the United Nations instruments at the centre of the report, the Commission harnessed the dams issue to a prominent body of international soft law.

The Commission's findings concerned the governance of development projects, writ large.

Looking back at the evidence in its Global Review of Large Dams, the Commission found that: "Governments, in constructing dams, have often found themselves in conflict with basic principles of good governance that have been articulated in the three international instruments [the United Nations covenants]. This situation still prevails today."[5]

Looking forward, the Commission proposed mechanisms to improve natural resources decision-making and prevent breaches of basic human rights from occurring in the context of dams again. The Commission presented a "Rights and Risks" framework to identify which stakeholders should be involved or represented in decision-making. Stakeholders would be identified based on whether they had a legitimate claim and entitlement (under law, constitution, or custom) that might be affected by a development project. No rights should automatically be considered superior to others. When rights of various stakeholders might overlap or conflict, good faith negotiations would be required to reconcile stakeholder interests.[6]

According to the report, the risks (or "loss of rights") of project affectees should be recognised and addressed in an explicit, open, and transparent fashion. Historically, the notion of risks had been applied to investors who risked financial capital on a project. The Commission broke new ground by highlighting the number of involuntary risk takers in both displaced and downstream communities as a result of dam building. Importantly, the Commission argued that the old-fashioned balance sheet approach that sought to trade off one person's loss against another's gain was unacceptable.[7] The Commission discussed some alternatives for providing water and energy services—

options that stakeholders might consider instead of large dams. However, the Commission's treatment of alternatives was fairly general.

Based on the United Nations instruments, the Commission distilled a set of core values for water and energy related decision-making, and seven Strategic Priorities. *(See Box 8.2.)* The Strategic Priorities formed the basis for 26 guidelines for options assessment and the planning, financing, building, maintenance, and decommissioning of large dams. The recommendations caution that conventional cost-benefit approaches to decision-making are insufficient but must be part of a richer, multi-criteria approach. They are a mixture of "carrots" that reward good practice, such as performance bonds, and "sticks" to punish poor practice, such as the call for five-yearly evaluations of dams' performance.[8]

> *The Commission grounded its recommendations in three United Nations instruments.*

The WCD was ahead of previous global public policy endeavours by siting human rights norms front and centre in the debate over large-scale infrastructure projects. But the WCD was not radical in this accomplishment either. As acknowledged in its final report, the international development discourse was already moving toward siting human rights at the centre of development more generally. The *UNDP Human Development Report 2000* shows how: "Human rights [constitute] the fundamental framework within which human development must be pursued."[9]

The human rights and right to development principles at the heart of *Dams and Development* were highlighted in the report's launch, in November 2001.[10] The keynote speaker, former South African President Nelson Mandela, addressed the role of dams in alleviating hunger. A spokesperson for Mary Robinson, UN Commissioner on Human Rights, lauded the WCD for its attention to human rights.[11]

The Commission Passes the Baton

At the launch of the Commission's report, Chairperson Kader Asmal presented the main findings and recommendations and declared the World Commission on Dams formally disbanded, its work complete. He turned the report over to the World Bank and IUCN and other concerned stakeholders for earnest consideration and adoption.

Box 8.2

The WCD's values and priorities

The WCD's Five Core Values:
equity, sustainability, efficiency, participatory decision-making, and accountability.

The WCD's Seven Strategic Priorities:
gaining public acceptance; comprehensive options assessment; addressing existing dams; sustaining rivers and livelihoods; recognising entitlements and sharing benefits; ensuring compliance; sharing rivers for peace, development, and security.

The full set of Strategic Priorities is summarised in Appendix 4.

Source: World Commission on Dams, Dams and Development: A New Framework for Decision-Making (London: Earthscan, 2000).

Professor Asmal and his fellow Commissioners saw the WCD Forum members as among the most important ambassadors for the WCD's principles and recommendations. Not only were these institutions influential players in the dams industry and the broader dams debate themselves—whose actions could change the nature of the dams business—but they were also opinion leaders. They were institutions and individuals with far-reaching networks whose views on *Dams and Development* could influence the conventional wisdom about the report.

Responses to the Final Report

The high stakes for diverse groups in the dams debate and the high expectations for the WCD report created a tense atmosphere for its launch and dissemination, and pressure for major industry and NGO players to respond. The tone of the initial responses ranged from glowing to scathing, with the majority being cautiously receptive. But one generalisation is possible about the report's reception. Institutions and individuals around the world were reading it closely and felt compelled to respond publicly. In the words of one Forum member, "People are poring over it."[12]

In fact, it was not possible for most institutions to provide official responses at once. The timing of official reactions reflected the nature of consultative procedures within institutions and networks and was also a function of the report's accessibility because of language and other factors. But even those organisations that did not make an immediate statement registered their interest in the report and their intention to form task forces to consider its implications.[13]

Because this assessment was completed only a few months after the release of the WCD report, it has been possible to capture only the first official statements and press coverage, and a sampling of individual comments.[14] Precisely because of the changes to business as usual required by the report and the time taken for dissemination and adoption, it will require years for the report to be read, interpreted, operationalised, tested, and evaluated around the world, and its full impact measured.

Non-governmental Organisations

A majority of international NGOs welcomed the WCD's final report and called for immediate action by bilateral and multilateral institutions to implement its recommendations. Their demand for change in financial institutions reflected NGOs' interpretation of these institutions' make-or-break roles in the project cycle of most large dams. It also reflected Northern NGOs' demand for accountability for their tax monies that contribute toward these public institutions. *(See Box 8.3.)*

NGOs generally supported the findings in the WCD's Global Review, for they had obtained the normative judgement on poorly conceived and planned dam projects that they had sought. They voiced approval for the strong emphasis on negotiation, due process, and justice for dam-affected people in the WCD's forward-looking framework and recommendations.

In their advocacy efforts, international NGOs tended to focus more on the content of the report on its own terms than on the process of the WCD. Their approach tended to be: "The process was not perfect, but the product was surprisingly good."[15] However, they did refer to the importance of the diverse stakeholder representatives on the Commission having reached consensus on the findings as a rationale for broader adoption and compliance.[16] Regional groupings of NGOs and movement leaders who had access to the English language publication echo this initial NGO statement. (Even several months after the WCD report's release, the summary had not been translated and fully disseminated and discussed at a grassroots level—*see "Peoples' Movements and Community-based Organisations," below.*)

A minority of advocacy NGOs and social movement leaders came out against the report. An article in the British press by one anti-dam activist was emblematic of this reaction. Philip Williams, the founder and former President of International Rivers Network, wrote that the WCD's failure to reject large dams technology altogether, by focussing instead on the weakness of decision-making processes, was "an unacceptable compromise for the global anti-dam movement."[17]

The report received an equally enthusiastic reaction from international environmental organisations, also among the first to respond. Such groups as IUCN and World Wide Fund for Nature welcomed the report's lengthy treatment of ecosystem management and recommendations for environmental flow releases. They supported the call for comprehensive and meaningful options assessment in water and energy planning.

IUCN's reaction bore a special significance, as it was one of the original convenors of the WCD as well as the world's largest conservation organisation. IUCN's Council passed a resolution at its Congress in early February 2001 to establish a task force to "define concrete avenues for implementation."[18] Later that month, it announced that its Water & Nature Initiative, a major follow-on from the World Water Vision process, was to incorporate the WCD's framework for decision-making and selected recommendations. The significance of IUCN's role in promoting and implementing the WCD recommendations was sealed when outgoing WCD Secretary-General Achim Steiner was named as the incoming Director General of IUCN.[19]

Peoples' Movements and Community-based Organisations

As of this writing, only a small number of statements have been issued from community and dam-affected peoples' groups because of language issues and logistical difficulties of diffusion. Translations of the WCD's summary report into Spanish, Portuguese, French, Japanese, Chinese, German, Russian, and Hindi had been released by

Box 8.3

International NGO responses

A Call to Action

- "... Public financial institutions should immediately and comprehensively adopt the recommendations of the [WCD] and should integrate them into their relevant policies... In particular, as recommended by the WCD, no project should proceed without the free, prior and informed consent of indigenous peoples, and without the demonstrable public acceptance of all those who would be affected by the project.
- All public financial institutions should immediately establish independent, transparent and participatory reviews of all their planned and ongoing dam projects...
- All institutions which share in the responsibility for the unresolved negative impacts of dams should immediately initiate a process to establish and fund mechanisms to provide reparations to affected communities that have suffered social, cultural, and economic harm as a result of dam projects.
- All public financial institutions should place a moratorium on funding the planning or construction of new dams until they can demonstrate that they have complied with the above measures."

—An International NGO Coalition[a]

"The following steps need to be taken:

- That governments and the private sector apply the criteria and good practice guidelines outlined in the report and publicly commit to undertake comprehensive options assessments before proceeding with the construction of any dam...
- That OECD countries publicly commit to not construct any further large dams at least for the next two decades...
- That all interest groups pledge to not enter into construction of mega-dams (i.e. those over 100 metres in height...) as the social, ecological, and financial evidence necessitates a worldwide moratorium on such dams."

—WWF International Position Statement, February 2001

"The work of the Commission represents a fair and balanced assessment of both benefits and costs, with input from all constituencies through high-quality reviews, public hearings and thorough information gathering. The Commission has created a knowledge base that goes beyond what any individual organisation could possibly have compiled.

What the WCD has finally given to us is not a final verdict on dams. But it has opened up a new path, a new approach to build upon. As such, the Report of the Commission forms a landmark in the history of the development and operations of dams."

—IUCN Position Statement, February 2001

"The Commission, evading its main task of adjudicating the 'development effectiveness' of dams, emphasises that it is poor planning of past dams that has caused unnecessary harm. This contradicts critics' charges that it is the dams themselves, no matter how well planned, that inevitably create unmitigated social and ecological impacts... The World Commission on Dams was no 'truth commission,' but more of a 'peace process.'"

—Philip Williams, anti-dam activist[b]

a Excerpted from "A Call to Action," by International Rivers Network and the Berne Declaration, with 109 additional signatories from around the world, 16 November 2000.

b Philip Williams, "Lies, Dam Lies," The Guardian (22 November 2000).

mid-2001. However, much of the process of discussion and mobilisation at the community level was just beginning.

The first statements, from dam-affected people in Brazil, Canada, Southern Africa, Nepal, and India, indicate that they found much in the WCD report to hearten them.[20] *(See Box 8.4.)* In particular, the Commission's recognition of the injustices perpetrated upon many displaced and project-affected people in the past was seen as a vindication of their struggles. These dam-affected people seized upon the report's call for review of problem projects and also its call for obtaining the free, prior, and informed consent of indigenous and tribal communities before a dam is built. If anything, dam-affected peoples' groups expressed disappointment that the Commission did not call for the free, prior, and informed consent of all people to be displaced by a reservoir, but instead used the less precise notion of "public acceptability."[21]

Box 8.4

Peoples' movements and community groups' responses

"The era of large dams and its [sic] grievous social and environmental impacts has come to an end. The era of dams built against the will of dam-affected people has come to an end. Brazil needs to move forward to this era. No new dams should be built in Brazil without the 'demonstrable public acceptance' of dam-affected people."

—Brazilian Movement of Dam-Affected People[a]

"As a direct result of the inequity between those communities who pay the costs of large dams and those who benefit from these same dams, we strongly urge for affected communities to be direct beneficiaries of existing dam projects."

—South African communities and non-governmental organisations[b]

"As peoples who have been dispossessed and devastated by the adverse biophysical, socio-economic and cultural effects of water and energy projects, we call upon international financial institutions to refuse funding to all water and energy projects for which the consent of the peoples or communities affected has not been obtained."

—James Bay Cree Nation and the Pimicikamak Cree Nation[c]

---

a Brazilian Movement of Dam-Affected People,"The Brazilian Movement of Dam-affected People (MAB) and the World Commission on Dams (WCD),"9 February 2001.

b Southern African communities and non-governmental organisations,"Southern African Call to Action,"23 November 2000.

c James Bay Cree Nation and the Pimicikamak Cree Nation, "Statement of the on the occasion of the release of the World Commission on Dams final report," Undated.

At a more profound level, dam-affected people and their supporters had hoped that the spirit of Commissioner Medha Patkar's "comment" to the WCD report[22] would have underpinned the entire report. In her note, Ms. Patkar calls dams "a symptom of the larger failure of the unjust and destructive dominant development model… Addressing these issues is essential in any attempt to reach an adequate analysis of the basic systemic changes needed to achieve equitable and sustainable development and to give a pointer towards challenging the forces that lead to the marginalisation of a majority through the imposition of unjust technologies like large dams."[23] For example, the Brazilian Movement of Dam-affected People (MAB) lamented that the WCD had failed to go beyond the recognition of dams' economic, social, and environmental problems "to unmask the private interests moving the dam industry around the world."[24]

In spite of these reservations, dam-affected and community-based organisations have indicated that the WCD report is something they can work with. In two of many such examples, MAB has called on Brazilian national stakeholders to convene a national commission on dams—in the spirit of the WCD—to assess Brazilian dam performance and planning and address outstanding reparations issues. The WCD report has invigorated a range of community-based organisations and NGOs in Kenya, which have begun pressing for its recommendations to be implemented in current dam projects in their country, especially as regards stakeholder participation in options assessment and project planning. *(See Box 8.5.)*

Multilateral Development Banks

The World Bank's reaction to the WCD report was intensely anticipated by a wide range of stakeholders, not least because it was a convenor of the Commission. By the time the WCD report was released the World Bank was involved in only 1 percent of large dam-building internationally[25] but its agenda-setting power in development discourse and its continuing leverage with client governments gave it considerable opportunity to influence the "international acceptability" of the WCD's recommendations.

The initial response from President James Wolfensohn at the London launch of the WCD report was warm and congratulatory. He was non-committal about the likelihood of the World Bank's adopting the recommendations. He emphasised that the report must be debated by the World Bank's Board and client governments before defining a way forward. But he nonetheless suggested that the World Bank would find many opportunities for adoption of the recommendations.

As months passed and the World Bank's internal evaluation progressed, its position seemed to harden against taking major action. A "progress report" delivered by Senior Water Advisor John Briscoe at the WCD's third Forum meeting indicated that the prospects for significant change

Box 8.5

Responses at a country level: Kenya

"Over the years the majority of Kenyan contenders in the dams debate were just whispering. Having read the Commission's report we are convinced that the Kenyan stakeholders had some things to tell the Commission regarding the experiences of dams in the country. Fortunately the Commission listened. The Commission told its story. And that has made all the difference. The various interest groups in Kenya are more than whispering now; they are dashing fully into voice. For this [hope] to be realised there is urgent need to develop a clear policy framework on stakeholder participation and mechanisms that not only facilitate but also ensure involvement of stakeholders.

The success of the WCD report in Kenya also largely depends on whether the policymakers will give the dams debate top priority; actively promote an enabling environment through adopting legal, political, economic, social, financial, and economic measures, etc. Besides, the country's development partners such as the donor agencies will have to demonstrate their commitment to the principles captured in the WCD report...

Due to the rising power of NGOs on the local scene, the influence and impact of Dams and Development in the decision-making process has a great impetus...The fact that the WCD has shifted the centre of gravity in the dams debate to focus on options assessment and participatory decision-making has had ripples in the implementation of proposed dam projects in Kenya."

Source: Excerpted from Josphat Ayamunda, "Dams and Development, Kenyan Perspectives on the World Commission on Dams." Draft Research Report for the WCD Assessment, January 2001.

within the Bank were fairly remote. The World Bank evaluation team, led by Dr. Briscoe and a senior official of the division for Environmentally and Socially Sustainable Development, had toured seven client countries to harvest official reactions to the WCD report.[26] As related in Dr. Briscoe's statement, World Bank contacts in Southern governments approved the core values of the WCD report but balked at the specific recommendations, demanding no new conditionalities on dam building. *(See "Governments Agencies," below.)*

Based on this feedback, the World Bank's evaluation team proposed a modest suite of follow-up activities. The World Bank said it would create a resource unit to gather information about good dams practice in line with the WCD recommendations and to "consider" how the WCD report might inform the World Bank's water resource sector strategy, then under development.[27] The significance of these steps was difficult to assess at the time, since each one had the potential to develop into a strong follow-up to the WCD or to wither away. But the overall message was that the World Bank would rely upon the interest of individual project managers in the institution and client governments to demand the skills and information to pursue the WCD's recommendations. Without the political will of senior World Bank officials to operationalise the recommendations, there is no reason to assume that, by themselves, these actors would have the incentive to do so. The burden for promoting compliance would fall once more to the World Bank's civil society monitors.

There were several ironic twists in these developments. The first lay in the World Bank's use of Southern government reservations as a rationale for its unenthusiastic response. When it convened the Gland meeting and supported the WCD process, the World Bank had focussed the stakeholder identification exercise on international interest groups in the dams debate, rather than focussing on its client governments. In other words, the World Bank had gone to the international NGOs, academic experts, businesspeople, and quasi-governmental development authorities to seek ways to break the deadlock in the dams debate. However, in the latter stages of the WCD's process and in the post-launch evaluation, the World Bank stated that client governments' responses to the WCD report would be the "acid test."[28]

There was also, to the outside world, an opaque relationship between the World Bank and its clients regarding who set the agenda. The World Bank sought guidance from client governments on how it should move forward. It readily accepted their proposals for no new conditionalities, although in the past it had not been shy about imposing multiple conditionalities. Simultaneously, these governments emphasised their appreciation of the World Bank as an honest broker in dam-related dialogue, but only, apparently, if national sovereignty remained supreme.[29] The decision to make few changes to business as usual apparently suited both sides.

The World Bank's response initiated criticism from an international coalition of NGOs and social movements that stated in an open letter to President Wolfensohn: "If the World Bank does now not feel committed to this [the WCD's] consensus, it indicates that the multi-stakeholder approach was not meant to effectively resolve the problems which brought about its creation, but to deflect opposition or to buy time. If the World Bank does not effectively adopt and implement the WCD's recommendations, NGOs may be less inclined to engage in future multi-stakeholder dialogues with the World Bank."[30] These developments coincided with the World Bank's revision of its resettlement policy, a policy with obvious implications for the dams arena. The revised policy drew fire from an even broader cross-section of international NGOs and social movements for lack of clarity, and for failing to learn from lessons of the past decade.[31]

The World Bank's response elicited widespread criticism from civil society.

The World Bank's response contrasted with that of the Asian Development Bank (ADB). The ADB held a meeting in February 2001 with government representatives from "countries with substantial hydro resources for water and energy uses" as well as development consultants, NGOs, and regional institutions to discuss the WCD report and implementation issues.[32] The ADB compared its own internal review of large dams in Southeast Asia with the WCD's findings and found the economic, technical, social, and environmental failures of large dams to be largely similar. Although the ADB's consultation with its client governments showed there was a long way to go before governments would adopt the spirit of the WCD's findings,[33] the ADB nonetheless told WCD Forum members how it intended to close the gaps between its existing guidelines and the WCD's recommendations.[34] At the final Forum meeting, the ADB called upon Forum members to enter into a good faith dialogue about the responsibility of multilateral development banks to address compensation issues.[35]

The WCD report also received a spirited response from the African Development Bank (AfDB), although the AfDB had not undertaken a consultative process like the ADB's as of this writing. The AfDB's President wrote to Professor Kader Asmal congratulating him on the Commission's achievement and stating that the AfDB intended to incorporate the WCD's recommendations in its own Integrated Water Resources Management Guidelines.[36] *(See Box 8.6.)*

Export Credit Agencies and Bilateral Aid Agencies

The Commission had foreseen a strategic role for international financial institutions, particularly for export credit agencies (ECAs), as levers in the dam-building process. The WCD Secretary-General had made a special presentation to the OECD working group on ECAs, and the WCD report was subsequently taken up for discussion at that forum.

In the United States, the Export-Import Bank was quick to scrutinise the WCD report and incorpo-

Box 8.6

Multilateral development bank responses

"Once we 'get the elephant out of the room' (no new conditionalities) then there are a host of ways in which countries are anxious to engage with the many good ideas in the WCD report, and to work with the World Bank (and others) in improving practice.

How the Bank plans to build on the WCD Report? The Bank will use it as a valuable reference to inform its decision-making process when considering projects that involve dams. The Bank will continue to support dams that are economically well justified and environmentally and socially sound."

—John Briscoe, World Bank[a]

"We feel that this report represents a major milestone in the assessment of economic, technical, and environmental performance of large dams. The report has been based on an exhaustive review involving broad stakeholder participation resulting in sound conclusions and recommendations."

—Oumar Aw, African Development Bank[b]

"The WCD Report provides a roadmap to move from the present, often-unsatisfactory process for planning, design, construction, and operation of dams, to a more equitable and sustainable one."

—Preben Nielsen, Asian Development Bank[c]

[a] Statement at the third Forum meeting, 25–27 February 2001.

[b] Letter to Prof. Kader Asmal from Oumar Aw, African Development Bank President, 26 January 2001.

[c] Statement at the third Forum meeting, 25–27 February 2001.

rate elements of its recommendations into draft guidelines for environmental and social impact assessment.[37] However, progress toward actual reform of the guidelines was somewhat dependent on ongoing discussions with other OECD export credit agencies, as the United States and its counterparts wished to keep their industries on an even, competitive keel.

Many bilateral agencies were actually represented on the WCD Forum or had made financial contributions to the WCD. Therefore, they demonstrated a sense of significant investment in the process and subsequent interest in implementing the final report.[38] Donor representatives from Switzerland, Sweden, the United Kingdom, Germany, and Norway attended the third WCD Forum meeting. At this meeting, they were uncharacteristically forthright in expressing their support of the Commission's report and their views on appropriate follow-up mechanisms for dissemination, compared to their modest profile during the WCD process itself.

Box 8.7

Bilateral agency and export credit agency responses

"We plan to use elements of the report as guidance in our guidelines. We don't advocate endorsing the report completely or ditching it completely because of one flaw... It'll be recognised for years to come. People will build on it."

—Official of the U.S. Export-Import Bank, speaking in his personal capacity[a]

"The final report of the Commission contributes to rational discourse on large dam projects."

—Heidemarie Wieczorek-Zeul, Minister for Development Co-operation, Germany[b]

[a] Interview with U.S. Export-Import Bank senior official, 7 March 2001.

[b] BMZ—Bundesministerium für wirtschaftliche Zusammenarbeit und Entwicklung, "Wieczorek-Zeul begrüßt Bericht der Weltkommission 'Staudammprojekte,'" Press release, Berlin, January 2001. Authors' translation from the German.

Bilateral aid donors received the WCD report with concerns similar to Southern government officials: Would the WCD recommendations be compatible with their existing policies and regulatory frameworks? However, given that most new dams will be built in the South, Northern officials tended to articulate a more flexible view about how the recommendations would be used. *(See Box 8.7.)* They emphasised the complexity of their existing guidelines and typically voiced support for adapting and adopting the WCD recommendations as appropriate.[39] An official of Britain's aid agency said: "The process is devalued if there is no discussion and debate. It goes back to the basis on which we were supporting the WCD (in the first place)—global public policy that supports national policies."[40]

United Nations Agencies

United Nations agencies provided a warm response to the WCD report *(see Box 8.8)*, which is significant given that the WCD report places a set of UN norms at the heart of its analysis and forward-looking framework. The UN agencies, many of which had partnerships of some kind with the WCD during its process, expressed appreciation for the usefulness of the WCD's framework to all types of development, not just dams. Overall, their approach was constructive and indicated a willingness to promote the recommendations in their ongoing development work and in their dialogues with governments. The head of the United Nations Environment Programme (UNEP), Dr. Klaus Töpfer, both welcomed the report's contribution to development debates and offered to host the WCD's follow-up body—the Dams and Development Unit—in UNEP offices, to facilitate dissemination to government stakeholders.[41] The World Health Organization (WHO) praised the WCD report for acknowledging the myriad and often complex effects of dam-building on public health and recognised the rights-and-risk framework as a "leap forward in development planning" overall.[42] The Food and Agriculture Organization of the United Nations (FAO) faulted the WCD for under-stating food security concerns but promised to carry forward the recommendations in a forthcoming international multi-stakeholder dialogue on Water, Food, and the Environment.[43]

International Industry and Trade Associations

International industry, which had proved the most organised segment of the private sector in their involvement with the WCD, provided a mixed response to the final report. *(See Box 8.9.)* Industry had sought, above all, clearer rules of engagement for its involvement in dam-related projects.

Box 8.8

United Nations agency responses

"The [WCD] report deserves a strong endorsement by the relevant UN specialised agencies... It has laid the foundations for a new approach to development in the coming decade, taking the Rio principles beyond their original scope into a more comprehensive and more participatory framework. If the report meets with broad support from all development stakeholders, then the scene is set for truly sustainable development in the 21st century."

—World Health Organisation[a]

"Large dams are required in some cases. We cannot afford to disregard any option to increase food supply in food deficit countries. But decisions to build dams must be taken in a responsible manner. It is in this spirit that we welcome the report of the WCD. We understand it as a framework for responsible decision-making, not as a verdict on dams."

—Food and Agriculture Organization[b]

"Controversy centring around the construction and operation of dams shows that, although the concept of sustainable development aims at balancing socio-economic development and environmental management, the international community is yet to develop an appropriate policy framework for reaching the goals of sustainable development. Equity, governance, efficiency, transparency and accountability based on open dialogue among all stakeholders can only be furthered by a serious consideration of the Commission's findings by all policy-makers."

—Klaus Töpfer, United Nations Environment Programme[c]

---

a World Health Organization, "Risks, Rights, and Negotiated Agreements." Response to the WCD's final report, 30 November 2000.

b Statement at the WCD's third Forum meeting, 25-27 February 2000.

c United Nations Environment Programme, "UNEP chief welcomes new report on impacts of dams as major contribution to future energy and water resource policy-making." News Release No. 00/129, 17 November 2000.

Moreover, it hoped these new rules could be quickly implemented, so as to provide minimal interruption to projects—and not so rigorous as to halt projects altogether. Industry's main questions had been: Under what circumstances would it have *carte blanche* to proceed with dam projects? Under what conditions would proposed dam projects raise significant social or environmental problems that were bound to elicit civil society protest? Industry's colleagues in international financial institutions held similar priorities: they hoped the WCD would provide straightforward criteria, compliance with which would usher projects through the pipeline with minimal conflict.[44]

The majority of corporations and trade associations expressed disappointment that the WCD had not provided such a straightforward solution. They perceived that the WCD report introduced uncertainty in timing and outcomes to the water resources development process. In particular, they singled out the recommendations for stakeholder dialogue on options and full negotiation between dam developers and project-affected people as introducing risks and uncertainties for project developers.[45] Kader Asmal vehemently contested this notion; he argued that the Commission's recommendations were intended to reduce the high transaction costs that accompanied the current conflicts over dam projects.[46]

But even beyond the project level, it was clear the WCD's proposed measures in their entirety would require significant time to be taken up by the responsible parties—especially national governments—and translated into action. By identifying bad governance as the root of poorly conceived and implemented dam projects, the report cast to societies the ongoing challenge of reconsidering their decision-making procedures. Corporations had clear roles to take, but they were also reliant on slow-moving political processes. The host societies were to define in large part the appropriate modes of participation and negotiation. Given the range of stakes and responsibilities held by private companies in dams projects, the degree of involvement of international companies and investors in identifying and negotiating with other stakeholders, once dams were chosen as the preferred option, was not entirely clear.[47] This element of uncertainty perplexed all but the most progressive companies.

Industry members on the Forum were also discontented with the final stages of the WCD's process, for they felt the lack of consultation with Forum members during the process of report preparation had weakened the quality of the report.[48] As our discussion of the political and practical trade-offs of full transparency revealed *(see Chapter 7)*, it is highly unlikely that industry representatives would have settled for small editorial changes in the late stages. They would likely have sought far-reaching changes in the content of the report[49] and even, further research to change the tenor of the find-

Box 8.9

International industry and trade association responses

"The report itself is not balanced. The benefits of dams are largely under-estimated or simply ignored, particularly as regards electricity supply. Concerning affected people, the Report speaks of resettlement, but there is no mention of stabilisation of the lives of people by providing them with water and power. The authors make sweeping generalisations about deficiencies of dams, based on a very small sample of large dams."

—Raymond LaFitte, President, International Hydropower Association[a]

"We consider the WCD report simply as a useful document to generate further discussion, but absolutely inadequate, as it stands, to find the required sustainable solutions. We do not accept the unbalanced judgement on the role of existing dams. The 26 WCD guidelines as they currently stand are considered unrealistic for application."

—Felix Reinders, International Commission on Irrigation and Drainage[b]

"As Nelson Mandela stated in his London address, many people suffer from hunger, thirst, lack of running water, sanitation, and electricity. ICOLD believe [sic] that the WCD recommendations will create an unacceptable level of uncertainty to the development process. ICOLD fear that public and private developers and financial institutions will view these delays as too time consuming and costly, and will stop water and energy development entirely, thereby compounding the human suffering referred to by Mr Mandela... ICOLD favours a balanced approach to dam and project development, giving a stronger voice to affected people and communities... Each country should consider the WCD recommendations and the ICOLD guidelines. However, each country must also consider its prevailing conditions, traditions, laws and needs."

—C.V.J. Varma, President, International Commission on Large Dams, ICOLD[c]

a Letter to the Editor, Hydropower and Dams, Issue Six, 2000.

b Statement at the third Forum meeting, 25 February 2001.

c Open letter to Prof. Kader Asmal, 30 November 2000.

ings. Industry representatives on the Forum were also distressed by the handling of the launch and the "leak" of the final report (as described in Chapter 7). The leak had given NGOs an upper hand in the public relations stakes, while at the same time, industry groups were at a disadvantage when asked to respond to journalists' queries.

Industry representatives on the Forum also reflected, behind the scenes, on opportunities they had missed for skilled engagement with the WCD. They noted that industry had been late to recognise the significant impact the WCD could have on the dams business. Several companies were aware of the process at the time of the Gland meeting, but did not devote attention or resources until much later. As a result, they felt that they had been out-manoeuvred by NGOs and peoples' movements in their engagement with the Commission. Moreover, some industry representatives noted that, as a block, industry had been slow to co-ordinate in presenting its interests to the Commission. Industry representatives also regretted their misfortune in being represented on the Commission by the CEO of a company—Asea Brown Boveri (ABB)—that withdrew from the large dams business halfway through the WCD process.[50]

Industry had hoped for guidelines that would be easy to implement.

Indeed, as with NGOs and other concerned groups, the international dams industry was far from being a monolithic block with similar interests and comparative advantages. The different stakes of the companies and trade groups in the dams debate had provided a challenge to their co-ordination throughout the process. Likewise, the potential impact of the final report on their bottom lines, and their subsequent responses to the report, were mixed.

Energy services companies that were sufficiently large and flexible to switch out of hydropower or companies with only a small portion of their portfolio in dams tended to give a level reaction to the WCD report. For instance, one project manager for a global power company comprised of generation, distribution, and retail supply businesses reflected this position when he said, "we don't care if we do hydro or not, we just want to undertake good energy projects."[51] ABB was a model of such a multi-service transnational corporation. [52]

Some companies even expressed support for measures that rewarded efficiency, effectiveness and good performance, recognising an opportunity for them to claim the competitive edge. For instance, Sweden's Skanska Corporation announced at the WCD launch that it would endeavour to follow the WCD's recommendations. "We find the Commission's work...represents a major stride for sustainable development, with open and transparent processes in which all affected parties can participate."[53] At least two corporations, the Nam Theun II Consortium (constructing a hydropower dam in Laos) and AES Corporation (constructing the Bujagali Hydropower Project in Uganda), evaluated their projects for WCD compliance after the report was launched.[54] They publicised these mostly favourable findings to Forum members and concerned NGOs. By contrast, engineering companies with vested interests in the production of dam-related technologies perceived the report's criticism of dams' past performance and its call for critical options assessment as fundamental threats to their business.

Professional dam industry associations provided a mixed, somewhat negative reaction to the WCD report. This was not surprising given that their institutional mandates called for the promotion of dams technology. In a statement, three international trade associations (the International Commission on Large Dams, ICOLD; the International Hydropower Association, IHA; and the International Committee on Irrigation and Drainage, ICID) expressed disappointment with the Global Review for failing to give adequate recognition to large dams' contribution to water supply, food security, and global energy needs.[55] Their main criticism was that the WCD had failed to compare dams' performance with the no-project option. In other words, the Commission had not analysed which development benefits would have been foregone without large dams. These critics were reluctant to cede conventional cost-benefit methodology to the normative framework adopted by the Commission.[56] It is worth noting that a significant proportion of ICOLD's membership comprises engineers and technocrats from dam-building government agencies in the South. In this regard, there has been significant overlap between the responses of ICOLD chapters to the WCD report, and Southern governments' separate official responses. *(See "Government Agencies," below).*

Government Agencies[57]

The WCD's recommendations anticipate a pivotal role for national governments in implementation. In the Commissioners' view, the good faith of all actors is required for negotiating acceptable outcomes. But above all, governments are required to create the enabling environment for such meaningful dialogues and interactions to occur. Furthermore, the consequences of adoption—or non-adoption—of the WCD's recommendations are of greatest import for the direct beneficiaries of water and energy development: societies and their governing institutions.

Many governments claimed their laws already followed the spirit of the WCD's recommendations.

The WCD had been initiated by the providers of international capital and their watchdogs under sustained pressure from transnational civil society and local social movements. The WCD's recommendations included a call for governments and their citizens to determine when and under what terms big capital should be mobilised for large dam projects. For all of these reasons, the reaction of governments—and Southern governments in particular—to the WCD was closely watched.

As with other sectors or interest groups, it is difficult to generalise about the responses by Southern governments to the WCD report. But perhaps more critically, it is important to distinguish the differences in response and motivation of the different ministries and interest groups within these governments and their bureaucracies. It would not be surprising, for instance, to learn that there were differences in reaction between environment and water ministries, or between departments for women's or indigenous and tribal people's affairs and finance ministries.[58]

Few official Southern government responses to the WCD report had been issued at the time of this writing. The sample size is small and it is biased toward water ministries whose comments have been the most widely disseminated. Often, these ministries are the agencies with a vested interest in the construction of new dams. These initial official comments do not necessarily reflect the richness of debate possible in legislatures and within and among political parties and congresses as the WCD report is further disseminated in these countries.

With these caveats in mind, the responses of Southern governments can be characterised as largely defensive in nature. This defensiveness stemmed partially from a misreading of the WCD's recommendations for governments. The considerable length and occasional lack of clarity in the report may have added to the confusion.

Several Southern governments, such as Brazil and Nepal, praised the core values of the WCD—equity, efficiency, participatory decision-making, sustainability, and accountability—and noted that these norms were already making their way into national policies and procedures.[59] The WCD was explicit that governments should begin national dialogues and review legal, policy, and institutional frameworks to establish opportunities for mainstreaming the WCD's recommendations. National governments nonetheless argued that they could not import the recommendations wholesale.[60] Indeed, many governments said that their existing laws and policies—such as requirements for environmental and social impact assessments—were already in line with the Commission's recommendations. *(See Box 8.10.)*

Southern officials based their opposition to the recommendations partly on national sovereignty issues and charges of Northern hypocrisy. They interpreted the Commission's highlighting of unmitigated environmental and social problems and dams' economic underperformance as an argument against future dams—a contravention of their perceived national imperatives. They found inadequate recognition in the report of dams' role in resolving food security, water, energy, and the overall development needs of their countries. They also suspected that the WCD had been an attempt by Northern governments and financing agencies to impose standards upon developing regions that the North had not followed when they had completed the majority of their dam building.[61]

Southern governments' objections underscored the long-time tensions over Southern dam building that had provided the very *raison d'etre* of the WCD. The WCD's objective had been to develop a new consensus around dam-related decision-making to break the costly deadlocks and bitter debates of the past and set the stakeholder groups off on a more constructive footing. By asserting that their policies were consistent with the WCD's recommendations, these governments denied the gaps between policy and practice that had formed a significant cause of friction in the first place. Furthermore, suspicions of the WCD's being driven by a Northern agenda belied the spirited contribution that Southern dam-affected peoples' movements had provided to many aspects of the WCD's process throughout its life, as well as the 50 percent representation of Southern stakeholders on the Commission itself and leadership of the Commission by a Southern government minister.

The governments of India, China, Nepal, and Ethiopia based much of their criticism on complaints about the WCD's process and methodology

Box 8.10

Government agencies' responses

"In Nepal's case, most of the procedures put forward by WCD are already in place in the form of several acts and rules. Adopting a new set of guidelines as mentioned in the WCD report, with its contradictory statements and yet to be polished prescriptions, would create confusion and chaos."

—Government of Nepal[a]

"The Commission's 'data base' is questionable, it is misleading... Even while applying the concepts of equity and participatory decision making, WCD has restricted its attention only to the groups which are adversely affected by a dam. It has failed to appreciate that there are much larger sections of society for whom the dam and the water supply flowing from the dam are nothing short of a life line... WCD's obsessive concern for preserving the rights of affected local people makes it distrust the entire public set up, even the legal framework of the country to which these people belong."

—Government of India[b]

"Nowadays the need for proper compensation and the rights of the people for development and decision[making] are enshrined in the national constitution and social and environmental impact assessments are a must in any small and large project implementation."

—Government of Ethiopia[c]

[a] Government of Nepal, "Preliminary Official Comments on the WCD Report." Submitted to the World Bank, February 2000.

[b] Government of India, Central Water Commission, "Final Report of the World Commission on Dams, Comments by Mr Gopalakrishnan, WCD Forum Member," February 2001.

[c] Government of Ethiopia, "Comment on Dams and Development... The report of the World Commission on Dams, Country Comment: Ethiopia." Submitted to the World Bank, February 2001.

(the contours of which we have described in Chapters 4 and 5). For instance, Ethiopia criticised the WCD for choosing too few dams for its sample and for selecting case study dams that were too old and did not incorporate new practices implemented in their regions in the 1980s and 1990s.[62] Ethiopia charged that the picture provided by the WCD failed to reflect the dynamism of evolving best practices in large dam building of the past two decades. The Government of India complained that official data on large dams provided to the WCD was not acknowledged in the WCD's studies or final report.[63] The Indian government noted that a global commission such as this should have undertaken far more intensive negotiations with such dam-building "giants" as India, China, and the United States in the early planning stages to avoid such missteps in the later process. In other words, early reactions from Southern governments explicitly faulted the WCD for a lack of credible process. The roots of this unease also lie in the inherent inability of a multi-stakeholder process, such as the WCD, to represent all viewpoints within and among governments.

Based on their unease with the WCD's process and product, some Southern government officials first took an "all or nothing" approach. They rejected the whole report on the basis of one statement or guideline that they considered incompatible with their circumstances. Such was the response of China, which called the WCD's emphasis on negotiated outcomes contrary to its established, and more democratic, decision-making procedures.[64] A Forum member seemed to articulate the Southern dilemma well when he said, "There's a fundamental tension between: Do we take the WCD report as is or do we move into dialogue, move toward better practice? There's an issue around going to the lowest common denominator to get any dialogue."[65]

However, this proclivity to reject the recommendations outright was already tempered in discussions among Southern government and NGO participants at the third Forum meeting of the WCD.[66] Participants agreed that policy reviews at the national level were necessary. Policy reviews would establish where governments were in compliance with WCD's recommendations and where they were not. They would provide an opportunity to identify gaps and form the basis of discussions for moving forward. Hence, a slow progression occurred in the dialogue within just the first few months of the report's dissemination, indicating that the Southern governments initial responses might be moderated by time, reflection, and the slow evolution and adoption of the new norms by countless other actors. In one example, the South African Department of Water Affairs and Forestry co-operated in mid-2001 in a joint congress with IUCN, the South African Committee on Large Dams, and the Environmental Monitoring Group (an NGO) to put in place concrete measures for bringing forward the WCD's recommendations in the South African context.[67]

In the North, governments tended to respond to the WCD report in the context of their development aid and export guarantee activities, not in their roles as domestic dam-builders and operators. Norway was an exception. The Government of Norway praised the WCD report but cautioned that it may have "gone too far in the direction of consensus-based decision-making systems." It noted that when weighing different development needs, decision-making can often end in disagreement and that the "superior competent authority"—in the case of dams, the Norwegian legislature—should make final decisions on behalf of the community as a whole. As with other national governments, Norway noted how many of its existing policy and legal frameworks already accommodated the WCD's recommendations for social and environmental impact assessment. Norway also referenced the body of international law on indigenous peoples' rights that it observes.[68]

The Third Forum Meeting and Institutional Follow-up to the WCD

A final meeting of the WCD Forum took place three months after the report's launch, for stakeholders to share their reactions to the report and plot a course for following up on the WCD. As the Commission had already disbanded, the Forum convened in the role of a loose decision-making body, much as the original Reference Group had done in Gland in 1997.

The meeting highlighted the divergent opinions about the WCD report in the stakeholder community. A minority of hard-line dam proponents declined to discuss follow-up to the WCD at all, but the majority of Forum members expressed their organisations' willingness to adopt the recommendations in some degree. The Forum meeting made clear that although dams remain a bitterly contested issue, the WCD had created a precedent for opposing parties to begin a dialogue and had delivered a report that provided a platform for future discussion. The facilitator of the Forum meeting[69] remarked that the level of dialogue that took place at this meeting would have been unimaginable only three years earlier.[70]

With few exceptions, meeting participants recognised that some form of institutionalised follow-up to the WCD was required at a global level to facilitate dissemination of *Dams and Development*. Members agreed that a small Dams and Development Unit (DDU) should be set up to replace the WCD Secretariat and to operate for at least one or two years. The future Unit would place a premium on reaching out to governments, and to support this goal it would be hosted by the United Nations Environment Programme.

Representatives from the World Bank, IUCN, International Rivers Network, the Lesotho Highlands Development Project, Struggle to Save the Narmada River, and Harza Engineering volunteered to act as a steering group to oversee the establishment of the DDU. They were voted in to this role under considerable time pressure. This hasty arrangement caused discomfort among members of the larger Forum,[71] who had differing views on whether and how the WCD's findings should be promoted. They saw the potential influence of the WCD report as being tied to the potency and mandate of the DDU, and hence under the full control of the steering group. Members from the larger Forum argued strongly for greater Southern government involvement in the WCD's follow-up activities.[72] In a concession to this point, the steering group pledged to add a UNEP representative to better reflect governmental interests.

Several aspects of the follow-up activity to the WCD constituted tacit acknowledgement of the process' strengths and weaknesses. The emphasis on reaching out to governments and the explicit harnessing of the DDU to a UN agency were acknowledgements that the WCD process had not involved governments as much as it should have. The appointment of a steering committee that broadly represented the political and sectoral categories on the Forum was a recognition of the general acceptability of the Forum's composition.

Conclusions

In the short term, Commissioner consensus did not translate into a broader stakeholder consensus. Indeed, initial reactions suggest a hardening of the positions that existed before the WCD. However, a closer look at these reactions suggests stakeholder willingness to grapple with the report, compare recommendations to existing policies and situations on the ground, and potentially put in place some ideas embedded in the report. If broad consensus lies in the future of the dams debate, it will be forged through a longer-term process initiated by, rather than concluded by, the WCD. As the WCD itself recognised, "...all concerned parties must stay together if we are to resolve the issues surrounding water and energy resources development. It is a process with multiple heirs and no clear arbiter."[73]

In the short term, Commissioner consensus did not translate into broad stakeholder consensus.

Could an immediate consensus among all stakeholder groups feasibly have been forged? The reactions to the report provide a basis for extrapolating what each group might have looked for in such a consensus. Based on their reactions, NGOs and social movements would likely have sought a more direct indictment of broader development processes. Industry groups would likely have rejected any articulation of a rights and risks framework that empowered affected communities to negotiate with industry on a time-consuming case-by-case basis. Various governments might well have focussed on issues of specific relevance to their national circumstance. Had the government and industry views prevailed, NGOs and social movements might not have continued their engagement with the process at all.

Hence, it is likely that an immediate, broad consensus among all stakeholders would not have been a viable goal. If anything, a process of stakeholder negotiation over the content of the WCD report might well have produced a report that only moved incrementally beyond the status quo. Arguably, such a report would have had a greater chance of being adopted wholesale by multilateral institutions, governments, and industry in the short term. However, it would almost certainly have lacked the support of NGOs and the social movements, and might have inspired even greater citizen protest. By focussing on forging a consensus among a smaller number of Commissioners, the WCD has produced a more aspirational text, but one at which dam-building nations and industries have balked. The promise for implementation depends largely upon ongoing constructive engagement by civil society groups with governments, international agencies, and the private sector, and the growing democratisation of national and global politics.

Endnotes

1. There were originally twelve members on the World Commission on Dams, plus one non-voting member, Secretary-General Achim Steiner. Halfway through the process Commissioner Shen Guoyi of China withdrew, and her employer, the Chinese Ministry of Water Resources, declined to provide a replacement. Hence, there were only 11 voting members. Achim Steiner's status on the Commission grew over time, and he subsequently signed the final report with the status of full Commissioner.
2. World Commission on Dams, *Dams and Development: A New Framework for Decision-making* (London: Earthscan, 2000), Executive Summary, xxxi.
3. World Commission on Dams, 2000, xxxi.
4. World Commission on Dams, 2000, Commissioners' Foreword, ix.
5. World Commission on Dams, 2000, p. 202.
6. World Commission on Dams, 2000, pp. 205-6.
7. WCD press release, "World Commission on Dams Launches 'Landmark' Final Report," 16 November 2001. Online at: www.dams.org/press/ (28 September 2001). See also World Commission on Dams, 2000, pp. 208-9.
8. Interview with Commissioner, 19 March 2001.
9. World Commission on Dams, 2000, p. 203.
10. At London's Canary Wharf, 16 November 2000.
11. Message to the Commission from Mary Robinson, United Nations High Commissioner for Human Rights. Delivered by Mr. Ahmed Fawzi, London, 16 November 2000.
12. Interview with Forum member, 27 February 2001.
13. Such as the press statements by the International Hydropower Association and International Committee for Irrigation and Drainage at the London launch event; general statements made by World Bank President James Wolfensohn, who noted that the report had to be submitted to the member countries of the World Bank's Board for consideration.
14. For detailed reactions to the final report, see www.dams.org/report/followups.htm (28 September 2001). This site is updated on an ongoing basis, and includes new material that has been posted since the authors completed this assessment in May 2001. All of the responses described in this chapter are posted on www.dams.org/report/reaction.htm except as noted.
15. For instance, the headline to an article by Himanshu Thakkar in *Himal* magazine refers to *Dams and Development* as "a surprisingly refreshing consensus report." Himanshu Thakkar, "Large Dams under the Microscope," *Himal* (1 April 2001). Online at www.himalmag.com/apr2001/report.html (28 September 2001).
16. Himanshu Thakkar, "Large Dams under the Microscope," *Himal* (1 April 2001).
17. See Philip Williams, "Lies, Dam Lies," *The Guardian* (22 November 2000). Online at: http://society.guardian.co.uk/societyguardian/story/0,7843,400894,00.html (28 September 2001).
18. IUCN Position Statement, "Working with the WCD Report." Released at the third Forum meeting, February 2001.
19. IUCN Press Release and website, www.iucn.org, February-March 2001.
20. For example, Narmada Bachao Andolan (Struggle to Save the Narmada River), "World Commission on Dams Report vindicates unjustifiability of large dams," 20 November 2000. See also references in Box 8.4.
21. Personal communication with Brazilian academic and liaison with the Brazilian Movement of Dam-affected People, 8 February 2001.
22. World Commission on Dams, 2000, pp. 321-2.
23. World Commission on Dams, 2000, p. 321.
24. Brazilian Movement of Dam-Affected People, "The Brazilian Movement of Dam-affected People (MAB) and the World Commission on Dams (WCD)," 9 February 2001.
25. Interview with World Bank Senior Advisor, 8 March 2001. See also John Briscoe, "Responding to the WCD Report: A Progress Report from the World Bank." Presentation to the WCD's third Forum meeting, Cape Town, 25 February 2001. "Message #5: Great concern [by developing country governments] about the apparent exit (prior to the WCD) of the World Bank from the dam business..." Online at: www.dams.org/events/f3_wb.pdf (28 September 2001).
26. Briscoe, 2001.
27. Briscoe, 2001.
28. According to a personal communication with a senior World Bank official, 29 July 2001. The official stated that President Wolfensohn made the terms of this "acid test" clear in his April 2000 meeting with the Chairperson and Secretary-General.
29. According to John Briscoe's presentation; personal communication with a senior World Bank official, 29 July 2001.
30. Berne Declaration and South Asian Network on Dams, Rivers, and People letter to James Wolfensohn of 19 March 2001, signed by 85 additional groups.
31. See, for example, World Rainforest Movement, "New World Bank Resettlement Policy Is Flawed." Bulletin 43, February 2001. Online at: http://www.wrm.org.uy/bulletin/43/WB2.html (28 September 2001); Dana Clark, "World Bank Resettlement, The Legacy of Failure." Bank Information Center Issue Briefing, August 2001. Online at: www.bicusa.org/ptoc/pdf/clark_reset.pdf (28 September 2001).
32. ADB Regional Workshop: Dams and Development. ADB Headquarters, Manila, Philippines, 19-20 February 2001. Summary of Proceedings, Draft of 2 March 2001.
33. Ramaswamy R. Iyer, public letter to Professor Asmal on the proceedings of the ADB consultation with client governments in Manila, 22 February 2001.
34. "ADB's ongoing and planned responses to the WCD's strategic priorities, best practices, and institutional responses," Internal ADB draft, February 2001.
35. Preben Nielsen, speech at the WCD third Forum meeting at Cape Town, 25-27 February 2001.
36. Letter to Prof. Kader Asmal from Oumar Aw, African Development Bank President, 26 January 2001.

37. The WCD report launch coincided with the U.S. Export-Import Bank's requirement to revise its guidelines. The Environment Division, which had been represented on the WCD Forum, took the lead in incorporating elements of the WCD report. However, as of this writing, progress on the revisions was stalled by the Export-Import Bank's Board of Directors' decision-making process and talks with other OECD export credit agencies.
38. For example, interviews with British, Swiss, and Norwegian aid agency representatives, 28 February 2001.
39. For example, interview with U.S. Bureau of Reclamation official, January 2001.
40. Interview with aid official, 27 February 2001.
41. United Nations Environment Programme, "UNEP chief welcomes new report on impacts of dams as major contribution to future energy and water resource policy-making," News Release No. 00/129, 17 November 2000.
42. World Health Organization, "Risks, Rights, and Negotiated Agreements." Response to the WCD's final report 30 November 2000.
43. Food and Agriculture Organization of the UN. Statement at the third Forum meeting, 25-27 February 2000.
44. Interview with IFC officials, 27 September 1999.
45. Interview with industry representative on the Forum, 28 February 2001.
46. Opening and closing remarks by Prof. Kader Asmal at the WCD's third Forum meeting, Cape Town, 25 February 2001.
47. That, at least, was the view of several corporate executives, who said the Commission "has gone too far and also not far enough" in defining corporate obligations.
48. See, for example, the response of the International Hydropower Association, February 2001. "However, the overall tone of the report is undoubtedly negative as regards the role of reservoirs, which undermines the constructive elements. The opportunity was not taken to use the Forum as a sounding board ahead of publication of the final report, despite the claim in the overview that the Forum was consulted throughout the process, which could have helped set the right tone."
49. Personal communication with former Secretariat staff, August 2001.
50. Interview with industry representatives on the WCD Forum, 27 February 2001.
51. Interview with project manager of international energy corporation at the third Forum meeting, 26 February 2001.
52. For instance, ABB plans to increase its share in alternative energy technologies to US$1 billion in the next five years. Announcement at ABB news conference, London, 7 June 2001.
53. Public statement of Axel Wenblad, Vice President of Environmental Affairs, Skanska.
54. AES Nile Power, "Appreciation of the WCD Report." Presentation to the third Forum meeting, 25-27 February 2001.
55. Letter and accompanying statement of ICOLD President, IHA President, and ICID President to World Bank President James Wolfensohn, "Joint Major Comments of ICOLD, ICID, IHA on the Report of the World Commission on Dams," February 2001.
56. Such as Honorary ICOLD President Jan Veltrop, who served on the WCD. Also, the British Dams Society, a member of the International Hydropower Association and ICOLD, was praised by the WCD Secretariat for its serious commitment to dialogue throughout the WCD process.
57. We concentrate this section on governmental responses from Southern governments, as the vast majority of future dam building is planned for the countries of the South. Most reactions from Northern governments are given in the aid and export credit agency context, in the section on international financial institutions, above.
58. For instance, the contrast between the negative and dismissive reaction to the WCD report provided by the Minister of Industry and Handicrafts of Lao PDR at the Asian Development Bank's meeting to discuss the WCD report, February 2001; and the positive response to the report provided by staff of the Science, Technology, and Environment Agency of Lao PDR in consultations with the authors in April 2001.
59. For example, the Nepalese government stated: "We are proud to note that we have been able to adopt these standards of international norms in our development process." Government of Nepal, "Preliminary Official Comments on the WCD Report." Submitted to the World Bank, February 2000.
60. Such was the response of the Chinese and Nepalese governments, Government of Nepal, 2000. The Commission wrote that its guidelines were not intended to be a "blueprint," but instead the basis for discussions and serious revision of current practices by all actors. World Commission on Dams, 2000, p. 313.
61. For example, Government of India, Central Water Commission, "Final Report of the World Commission on Dams, Comments by Mr Gopalakrishnan, WCD Forum member," February 2001. Similar views were expressed by Indian government officials at the National Consultation on the World Commission on Dams Report, 24-25 May 2001, and the WCD India Meeting, 26 May 2001, as documented by Ramananda Wangkheirakpam and Lakshmi Rao in their background papers for the WCD Assessment.
62. Government of Ethiopia, "Comment on Dams and Development...The report of the World Commission on Dams, Country Comment: Ethiopia." Submitted to the World Bank, February 2001.
63. Government of India, Central Water Commission, February 2001.
64. Government of China, Ministry of Finance, "China position on the WCD report." Memo to the World Bank, 9 February 2001.
65. Interview with Forum member, 27 February 2001.
66. Cape Town, South Africa, 25-27 February 2001.
67. WCD press release, "South African Symposium Endorses WCD Recommendations," 24 July 2001. Online at: www.dams.org/press/ (28 September 2001).
68. Norwegian Ministry of Foreign Affairs, "Norway's Comments on the Report of the World Commission on Dams," June 2001.
69. Professor Anthony Dorcey, who had facilitated the original meeting of the stakeholders at Gland. Prof. Dorcey's "Facilitator's Sense of the Meeting," a useful

documentation of this last Forum meeting, is contained in a publication by the Secretariat entitled "Final WCD Forum: Report, Responses, Discussions, and Outcomes," May 2001.

70. Personal communication with Gland workshop organiser, 28 February 2001.
71. Interviews with Forum members, 27-28 February 2001.
72. Plenary session of the third Forum meeting, 27 February 2001.
73. World Commission on Dams, 2000, p. 319.

Chapter 9

Conclusion

Was the WCD a watershed in global governance? The WCD marked a departure from previous global commissions and multi-stakeholder processes in significant ways. Because of its origins in the international organising of dam-affected people and their supporters, the Commission included representatives of project-affected communities and indigenous peoples for the first time. Commissioners came from all points on the political spectrum of the dams debate, rather than from a broad middle. For the most part, they were active practitioners in international networks, rather than the eminent persons of past commissions. Not least, the Commission made an explicit commitment to incorporate good governance principles in its work, such as independence, inclusiveness, and transparency, as a way of creating an effective platform for dialogue. These features added up to a Commission structure and process that were path-breaking by historical standards.

A commission with these elements of structure and good process—full representation of relevant stakeholder groups, independence from external influence, transparency, and inclusiveness in the work programme—was meant to create opportunities for broad stakeholder participation and thus, a rich base of common knowledge. Good process was also intended to build constituencies for implementing the WCD's recommendations.

Did the Commission succeed in implementing the good process to which it aspired? If so, how important was good process to the commitment of interest groups to promote and implement the WCD's recommendations?

This assessment shows that despite some flaws, the WCD process was essentially robust, allowing for a wide range of perspectives to be represented and incorporated in the Commission's work. The Commission succeeded in persuading diverse stakeholders to contribute to its knowledge base, so that a broader than normal group of actors was heard. The Commission transcended, rather than reproduced, fractures among interest groups in the dams debate by producing a consensus report.

The consensus among diverse Commissioners and the emphasis on broad consultation and a transparent process gave moral authority to the final report and amplified the Commission's advisory voice. Indeed, interest groups around the world awaited the final report with expectation and reacted publicly to it, providing another tangible measure of how the process engaged stakeholders. Because of these accomplishments, the WCD's process set a new standard for multi-stakeholder processes.

However, the range of positive and negative reactions to the final report suggests that the consensus among Commissioners did not translate immediately to agreement among the broader stakeholder constituencies. Whether it does so will depend upon stakeholder willingness to continue dialogue on the Commission's findings and recommendations. Indeed, the WCD's model for change relies upon champions in financial institutions, industry, government, and civil society working to implement the recommendations.

The effectiveness of the WCD's strategy for change may be gauged only years after the report's release, when changes in water and energy planning and dam building can be measured. The WCD produced an aspirational document with challenges for implementation whose promise for change lies in the medium- to long-term. The degree of stakeholder engagement that we have documented during the WCD's process and in the immediate aftermath suggests that the process is worth emulating—and improving upon.

Given that the WCD's process was robust, this chapter reviews the major elements of successful process that could be replicated in future commissions. Where the process was flawed and undermined stakeholder confidence, we suggest corrections that could be designed into future efforts. The chapter concludes by considering the legacy of the WCD for future processes.

Establishing a Credible Process

Representation

As a result of the contentious selection process, the Commissioners chosen for the WCD were prominent individuals in NGO, social movement, and business networks, as well as in government agencies. In many cases, Commissioners were selected because they were perceived as being affiliated with distinct constituencies. Such a composition set it apart from past commissions that were composed of eminent persons. This new model proved successful in engaging large international stakeholder constituencies in the Commission's work.

The WCD's initiators had no obvious benchmark to use when departing from an eminent persons' model to compose a diverse multi-stakeholder commission. Rather, the selection of Commissioners was the result of political negotiation. The WCD's own process later generated a useful framework for identifying stakeholders based on rights and voluntary and involuntary risks. This framework will be relevant to the formation of future multi-stakeholder processes.

Involving stakeholders in the selection of Commissioners increased their confidence in the Commission and their willingness to participate in the subsequent process. It would have been impractical for all concerned stakeholders to learn about and participate in Commissioner selection. Instead, the WCD's formation involved an ad hoc group of participants from the 1997 Gland meeting—the World Bank-sponsored gathering that called for an international review of large dams. The involvement of this group helped gauge the political acceptability of the Commission and promote awareness of the WCD in participants' networks. This experience demonstrates the usefulness of broadening the selection process in future commissions.

The different levels of organisation and interest across stakeholder groups at this time influenced their participation in the process. When the WCD was formed, the anti-dam movement was relatively well organised, which translated into coherent demands for representation. Before and after the Gland meeting, civil society groups refused to be subsumed in a single stakeholder category. They argued that "civil society" was sufficiently diverse that it merited several stakeholder categories: indigenous peoples, non-indigenous project-affected people, public interest advocacy groups, and environmental groups. In particular, civil society groups pressed for affected peoples' representatives and indigenous peoples' representatives to have their own seats at the table. This demand—which formed the basis for the Commission's political acceptability to the anti-dam movement—marked a departure from previous consultative and multi-stakeholder processes in which civil society slots were typically taken by NGOs close to the corridors of power in Washington, Nairobi, and Delhi.

By contrast, dam-building companies and utilities did not take an active interest in the WCD process in its early days, largely because they disregarded its significance. As a result, when private companies and utilities recognised the growing reputation of the WCD and became more active in the process later on, some felt under-represented on the Commission. Circumstances exacerbated this discontent—the Commissioner best positioned to represent corporate interests (Mr. Lindahl) slowly lost the confidence of industry groups when his company, Asea Brown Boveri, progressively withdrew from the large hydropower business. This gradual loss of representation made industry groups feel that they were losing ground in the debate and led them to establish a more formal industry network toward the end of the process. The contrasting experience of stakeholder groups suggests that stakeholders bear the considerable burden of defining constituencies and mobilising accordingly, if they are to feel adequately represented.

National governments were represented only modestly at the Gland meeting and in the Commission's formation—an outcome with considerable significance for the subsequent process. The Indian government would later point out that in its view, it and other major dam-building governments were excluded from the formative process. Despite the presence of a strong Chairperson and Vice-Chairperson from Southern governments, the WCD's ability to attract the political support of governments would become a serious issue in its later process and would inhibit their more enthusiastic engagement with the process and the final report.

The Commissioner selection process proved instrumental to stakeholders' willingness to engage in the WCD process.

The issue of government representation highlights one of the trade-offs that the WCD could not avoid. The muted participation of governments during the Commission's formation helped create the space for NGOs and social movements to participate. These groups believed that to involve governments integrally from the start would have delayed, if not stalled, the formation process.[1] Based on statements made by governments later in the process, their own involvement at the formation stage would likely have created a less broadly consultative process and a less aspirational outcome.[2] Hence, greater inclusion of governments would likely have led to the loss of civil society voices.[3] For example, the response of the Chinese government to the unfolding process certainly suggests that some governments were unwilling to sustain engagement with a broad spectrum of stakeholders. China's discomfort with the process triggered its rejection of the final report, which was significant given that almost half of the global population of large dams is located in China.

The option of having a Commissioner from the World Bank was never explicitly considered given the history of the dams debate and the WCD's genesis in civil society calls for an independent review. In the selection process, it was important to affirm the principle of diverse stakeholder representation, rather than to seek specific representatives from particular institutions or agencies. Indeed, representation of the World Bank on the Commission may have alienated social movements and NGOs. Such a development would have changed the entire character of the process and, likely, its results. In future processes, if the World Bank is more centrally engaged, even represented on a commission, it may place greater pressure on the institution to acknowledge ownership of the findings and recommendations. The price, however, would be diminished independence of the commission. Such potential trade-offs between representation and adoption of findings, on the one hand, and character of the process and results, on the other, are highly relevant to future processes.

Finally, the WCD experience suggests that adequate representation of stakeholders should extend beyond a commission to all the other organs of a process. As a sounding board, the WCD's advisory Forum was intended to capture diverse perspectives from the dams debate, and did so successfully. Secretariat diversity was important, because Secretariat staff members were the filters between the broad community of stakeholders and the Commissioners, and deployed their networks in developing the work programme. NGOs faulted the Secretariat for having no staff who had worked directly with displaced people. Industry groups criticised it for lacking technical dams expertise. According to the Secretariat, it was difficult to recruit senior staff with diverse sectoral and regional backgrounds because of relocation issues and the temporary nature of the assignment. Criticisms by Forum members suggest that quite aside from the performance of the WCD Secretariat, in a partisan arena each interest group hopes to see someone "like themselves" on a secretariat, and judges the legitimacy of a secretariat accordingly.

Women were well represented on the Commission itself, comprising five of the twelve original members. However, both the Secretariat and Forum had disproportionately small numbers of women or, perhaps more pertinent to issues of representation, they included few women or men who were sensitive to the gender-differentiated impacts of water and energy development, along with best practice in gender and development work. The discrepancy in numbers and the poor representation of gender advocates led women to feel marginalised in discussion forums. The WCD's final report might have had a stronger gender perspective running through it had there been more women and gender advocates in its Secretariat and Forum.[4] By not including more such voices, the WCD failed to meet its own standard for inclusiveness and neglected an important constituency.

In summary, representation of the full range of stakeholders, across government, business, and civil society as well as regions and disciplines, can considerably enhance the legitimacy of a multi-stakeholder process. In the case of the WCD, this potential was somewhat diminished by the lack of full industry confidence, largely because of industry's failure to mobilise early in the process and by the wariness of some governments. Yet, the alternative of relying exclusively on a middle ground, however eminent the representatives, would not have carried the same credibility with the range of people involved, and particularly not with civil society, whose calls for an independent review of dams led to the formation of the WCD. Hence, despite the trade-offs that make it impossible to satisfy all sides and despite the challenges of balancing various forms of representation, the potential legitimacy gains make the representative multi-stakeholder model worth emulating.

Independence

The WCD was born out of calls by civil society for an independent review of the global experience with large dams, with a particular focus on the role of international aid and credit agencies. Hence its independence—not only from funding agencies, but also from influence by various stakeholder groups—was a critical element of its legitimacy. At the same time, the success of the WCD relied on vigorous engagement by all stakeholders, so as to promote buy-in to the process and the final outcome. The simultaneous pursuit of independence and engagement certainly posed a challenge. Rather than seeking neutrality, the Commission sought balance in its engagement with stakeholders.

The Commission was independent from the convening institutions—the World Bank and IUCN—insofar as it was not answerable to them, these institutions were not represented on the Commission itself, and they did not control its operations or decision-making process. The WCD made a clear choice for independence over ownership by convening institutions, a choice that was critical to the Commission's legitimacy. The only debate in this regard concerned Secretariat staff's links with the convening institutions. In the WCD process, some industry stakeholders were concerned that the Secretary-General and three of ten senior advisers had strong prior ties with IUCN and that ecological issues would be given undue weight as compared to social or economic concerns. Although it is possible that the WCD's creators underestimated the role of staff members' backgrounds in building confidence, this example suggests that independence might be considered in the selection of secretariat members for future commissions.

The WCD's independence from convening institutions was critical to its legitimacy.

Maintaining independence by diversifying funding sources was a major accomplishment of the WCD that enhanced its broader legitimacy. The Commission explicitly sought financial support from government and multilateral agencies, the private sector, and civil society groups. This fundraising effort was time-consuming and overshadowed much of the work programme. However, the pay-off was worth the effort, for diverse funding sources demonstrated that the WCD was not beholden to any one set of interests. Indeed, it is a notable measure of success that Forum members and the general media did not criticise the WCD's funding strategy.

Also in the interests of independence, the WCD adopted a policy of only seeking money that came with no strings attached. This was more difficult to accomplish. In order to raise sufficient funds, the WCD did compromise this principle. For instance, the Commission accepted major donations (principally from bilateral and multilateral agencies) that were tied to specific events or studies. However, there is no evidence that these conditions forced the WCD to do what it otherwise would not have done, nor did they undermine the confidence of Forum members or other concerned stakeholders in the integrity of the process. Future processes will, similarly, have to handle such relationships cautiously to avoid donor influence.

Transparency

Transparency was central to the WCD's legitimacy for several reasons. Non-transparent decision-making processes in multilateral institutions and in large-scale development generally have been major causes of friction in the history of large dams. To mobilise broad input for its work programme, and therefore build credibility as a platform for dialogue, the WCD had to respond to stakeholder demands for transparency. Transparency was especially important because there were no formal accountability mechanisms between Commissioners and various constituencies. Hence, disclosing information about objectives, methods, and progress helped keep Commissioners honest to broader tides of opinion. Perhaps most important, the WCD had limited ability to facilitate broad consensus among contending interest groups during its lifetime. To leverage the WCD's influence in the dams debate, the commitment to transparency was necessary to disseminate new ways of thinking among these constituencies.[5]

The WCD did strive to, and substantially achieve, high standards of transparency. A widespread standard for transparency is that the convenors of policy processes communicate their objectives to relevant stakeholders in a timely manner, tell stakeholders how they can participate and how their inputs will be used, and communicate decisions in full.[6] The WCD publicised opportunities for participation in the work programme broadly. It disseminated terms of reference for the commissioned papers, and the draft and completed thematic reviews and case studies to all interested parties, and posted them on its award-winning website. In addition, stakeholder groups were engaged in the process by serving as reviewers of terms of reference and studies and by occasionally participating in meetings organised around the various studies.

The WCD's efforts to reach out in person to stakeholders and go beyond reliance on the Internet were important to stakeholders with limited Internet access. This included the majority of Southern stakeholders, even in elite institutions. Personal contact—through seminars, workshops, and official consultations—helped engage stakeholders and solicit their input more effectively.

The Commission's track record for transparency was tarnished toward the end of the process, however, when it did not communicate clearly whether the Forum would have an opportunity to review a synthesis of work programme results. The Secretariat was to compile a synthesis midway through the process based upon the myriad background studies that had been commissioned. Such a synthesis would have provided a succinct summary of the knowledge base that the Commission would use to prepare its findings and recommendations. Because of time pressures, this interim step was abandoned. Following the release of the report, the World Bank and stakeholders from the dams industry suggested that this lost opportunity for review compromised the technical merit of the report.

To build credibility as a platform for dialogue, the WCD had to respond to stakeholder demands for transparency.

Although discussion of *interim* findings compiled by the Secretariat based on the knowledge base may well have been constructive, disclosure of the Commissioners' draft *final* report, as some Forum groups desired, would have been counterproductive. A premature effort to build a broad consensus among stakeholders, via the Forum, might have risked undermining progress toward the Commissioners' consensus. Over two years, the Commissioners had developed a delicate internal dynamic based upon mutual respect and shared learning that did not exist among Forum members or the wider stakeholder community. Circulation of a draft for comment risked igniting politically charged debates among interest groups, which could have undermined Commissioner solidarity. The lesson is that the demand for transparency must be balanced with the often delicate dynamics of consensus among commissioners.

Another set of practical challenges to full transparency pertain to consultation in a global setting where stakeholders' use of information is limited by language and their access to information is limited by Internet availability. The transparency of the WCD's process was diluted for those non-English speaking stakeholders who could not understand the information. The WCD's record in translating information about the work programme from English into other languages was mixed. Although the final report itself was translated in full into Spanish and the summary into numerous languages, working documents were not translated. Because it is not practical to translate multiple drafts of working papers for stakeholder dissemination, a reasonable standard may be to translate essential framing documents and interim products into major world languages. Although translation and interpretation requires significant amounts of time and money, it should be an integral part of the time lines and budgets of future processes.

Finally, the Commission's efforts to disseminate information about opportunities for participation were not matched by its management capability to acknowledge stakeholder inputs once they were received. This is a problem that future processes can correct. For almost the first two years of the Commission's life, the Commission invited stakeholders to send written submissions on the development effectiveness of large dams. The process yielded a total of 970 submissions from institutions and individuals around the world and helped the Commission achieve an image of openness. Managerial problems somewhat undermined the mechanism's legitimacy. Consultants failed to integrate submissions and only in the late stages did the Secretariat have the resources to do so. Eventually, the submissions were sorted by theme and appended to thematic reviews after the full report was complete. They were included on a CD-ROM of the knowledge base that was mailed to stakeholders after the report's launch. However, the lack of early acknowledgement undermined the confidence of contributors that their submissions would be taken into account.

In summary, the Commission's mixed record on transparency demonstrates that it did not meet stakeholders' high expectations for information disclosure, although much of this could not have been managed without unrealistic increases in the length and cost of the process. However, the Commission also fell short in meeting some basic international norms of transparency, as in acknowledging stakeholder inputs or being completely clear about the role of the Forum in consultation. These experiences provide lessons for operational improvement that can be applied to future multi-stakeholder processes.

Inclusiveness

Given the diverse composition of the Commission itself, the WCD's knowledge-gathering process had to be inclusive of diverse viewpoints. Without an inclusive approach in the work programme, the Commission would not have held together.

The WCD's ability to create and maintain political space for diverse engagement rested in large part on its open-ended approach to knowledge gathering. Rather than defining criteria up front for the development effectiveness of large dams, the Commission invited stakeholders to present their own analytic and normative views of whether dam projects had advanced their society's development. The multidisciplinary case studies were in theory set up to elicit such converging and diverging views. This approach assured stakeholders that the process did not prejudge outcomes, and thus encouraged broad participation.

The WCD Forum demonstrated the inclusiveness of the process.

The WCD's regional consultations were important vehicles for the Commission to demonstrate its inclusive approach. These hearings, in South Asia, South America, Africa and the Middle East, and East and Southeast Asia, brought almost the entire Commission and Secretariat to Southern regions to reach out and listen to stakeholders. The Secretariat went to considerable lengths to include social, economic, and environmental topics, along with pro- and anti-dam perspectives on the panels. Payment for presenters' travel ensured that a range of presenters could attend, from community representatives to environmental experts to dam engineers to agency planners. Not only did these events raise awareness of the Commission's work during its process, but they were also a means of legitimising the process' outcome—the Commission could rest its report upon consultations with thousands of people.

Two important lessons from the regional consultations are relevant for future commissions and multi-stakeholder processes. First, even when meetings are carefully designed for balance and inclusiveness, the failed India meeting *(see Chapter 6)* serves as a reminder that the location and timing of public meetings is a political decision that can alienate stakeholders. When such decisions appear heavily biased toward one side or another, the commission risks destroying its ability to act as a convenor for broad stakeholder dialogue. In this case, the proposed meeting site appeared to bear upon a local controversy.

Second, the WCD succeeded in mobilising grassroots input for its hearings, which was notable for a global commission. WCD events often marked the first time that government officials had heard directly the voices of affected people and the alternative viewpoints of NGOs. This mobilisation owed something to the efforts of diverse Commissioners, Secretariat staff, and Forum members. But, in particular, grassroots mobilisation resulted from the efforts of a few highly co-ordinated, dedicated civil society groups who reached out to contacts at the community level with their own resources. Future processes will also rely heavily upon the networks of their staff, commissioners, and advisors to mobilise participation. Where such networks are limited in their reach, as they inevitably will be, it may be practical to assign additional resources to civil society groups and local actors to increase appropriate outreach. Vigorous outreach to local media to mobilise input to consultations would also be a cost-effective strategy in the future.

The WCD's advisory Forum also demonstrated the inclusiveness of the process by including organisations that had engaged in bitter wars of words and even physical clashes in the past over the legitimacy of dam projects. Export credit agencies that were backing controversial dam projects in the South joined the Forum alongside indigenous peoples' groups defending their ancestral lands from large dams. Large engineering firms that supplied dam equipment joined alongside civil society organisations that had arranged protests outside their corporate offices. Some Forum members refused to engage in direct dialogue with others, but many agency officials, community representatives, and NGOs came with—or developed—a listening ear. Although the effects are hard to measure, gathering such actors in the same room for three substantive Forum meetings was clearly an achievement.

While the act of convening such diverse parties was worthwhile and quite unusual, the ongoing engagement of Forum members in the WCD's work programme was sporadic and uneven. Structured opportunities for Forum members to provide guidance on the work programme were few. Commission and Secretariat members say they were informed and empowered by their two formal meetings with Forum members. However, the consultations fell far short of using the Forum as a "sounding board" for the direction of the WCD's final report as Forum members gained little sense of the Commission's internal deliberations. As a result, Forum members' ownership in the process and forthcoming product was quite tenuous. Between the second Forum meeting and the launch of the report, an intense aura of secrecy surrounded the final report's content, and the report surprised many Forum members upon its release. Many of these members were ill-prepared to receive and respond to the report.

The WCD experience suggests that advisory bodies have considerable value in providing a platform for exchange among conflicting interest groups. Such bodies can assist in furthering a commission's shared learning and advances in its members' thinking. For participants to reap tangible benefits from the experience, they not only need to be organised and motivated to participate themselves, but they also require regular updates about the progress of the work programme and the direction of a commission's thinking. The WCD excelled in providing Forum members with informational updates, but as our discussion about transparency indicated, they fell short of their own high standards in fully engaging the Forum.

The WCD managed to open up the knowledge-gathering process.

In summary, the WCD accomplished a process that was very inclusive by global standards. The Commission's insistence on welcoming all forms of evidence as a valid contribution to the knowledge base—the grassroots as well as the "official"—ensured that it was more democratic than technocratic. The effort to reach previously unheard voices also displeased some technical experts who were accustomed to being the dominant participants in such processes and this reaction will have ripple effects on future processes. The WCD's major achievement was that it developed sufficient authority as a convenor that it could create and strengthen the political space over two years of consultations to engage most concerned parties in the knowledge gathering process. Such inclusiveness provided a strong basis on which to rest the Commission's final report and sets high expectations for future multi-stakeholder processes.

The Legacy of the WCD

The Promise of a Representative Commission

The WCD reveals both the promise and the pitfalls of an advisory multi-stakeholder process. The promise is that selection of active practitioners can provide legitimacy with the full range of stakeholders engaged in a debate. The pitfalls are that determining representation within amorphous constituencies and expanding consensus among representative commissioners to a broad consensus remain challenges.

The WCD also provides lessons on how to support and promote the legitimacy of advisory commissions. The WCD experience suggests that if a multi-stakeholder process is to truly move beyond the divisive politics of an issue, representatives from the full spectrum of the debate must be at the table. The WCD provides a model where voices that have long protested decisions made about their lives in their absence can represent their views directly and share in developing a framework for future decision-making.

Who is a stakeholder? Who should be at the table? The WCD's report provides a framework for identifying who is a legitimate stakeholder, based upon rights and risks. This framework calls for full identification of the overlapping and intersecting rights of stakeholders in a country or river basin, combined with attention to who is taking voluntary and involuntary risks if the development project proceeds. The rights and risks approach could be used to identify legitimate stakeholders for dialogues in many development arenas—from the global to the national to the local. Based on legitimacy with a wide range of stakeholders, such bodies are well poised to act as "norm entrepreneurs," who articulate genuinely new formulations that, over time, diffuse and are accepted as new norms of conduct in the international arena.[7]

What Does Good Process Contribute?

In this assessment, we have examined the WCD process against the benchmarks of independence, transparency, and inclusiveness. Although the process did have flaws, we have concluded that it was essentially robust. As the stakeholder reactions above suggest, however, good process cannot by itself transcend divisive politics. Indeed, it would be naïve to suggest that it could. What, then, did attention to process bring to the WCD, and what does it promise for future processes?

The most significant contribution of good process is to support the legitimacy of a multi-stakeholder process. This is important because in contentious arenas, such as dams, not all differences can be reconciled through new information and cognitive advances. Ultimately, some differences are irreconcilable and will require a framework to decide which interests prevail. A legitimate process is an important defence against criticisms of this approach.

A good process can expand the range and variety of information and perspectives that feed into decision-making. The WCD's efforts at inclusion brought to the fore voices that have often been marginalised in the dams debate. The Commission cast a wide net, capturing the views of the displaced, along with the reports of consultants and the data banks of governments. This process enriched the knowledge base on which the WCD deliberated.

An important promise of a multi-stakeholder process is its ability to create a broader space for dialogue among stakeholders. The WCD proved only partially successful at this task. In their frequent face-to-face meetings, the Commissioners were able to transcend pre-conceived characterisations of other constituencies. The broader group of stakeholders had far fewer opportunities for interaction. Moreover, the regional consultations and, in large part, the Forum meetings were structured to inform the Commission, rather than as a two-way dialogue. Finally, the absence of an interim report that could stimulate a directed discussion among Forum members proved an obstacle to furthering stakeholder dialogue. Despite these design flaws, the WCD, nonetheless, did encourage far more communication across stakeholder groups than had occurred in the past years of the dams debate and additionally stimulated the formation of networks within stakeholder groups.

The Challenge of Implementation

Multi-stakeholder processes typically have little formal decision-making authority, and the WCD was no exception. Instead, multi-stakeholder processes are designed to win consent for implementation through a process of inclusion, with a particular focus on civil society and the private sector. A process structured around representative stakeholders holds the potential for genuinely new and transformative formulations that can break policy deadlocks—a contribution that is less likely to be achieved through governmental processes alone.

Representatives of the full spectrum of the debate must be at the table.

Yet, as the tentative and defensive reactions of Southern governments to the WCD suggest, a multi-stakeholder approach co-exists uneasily with the existing framework of international law based on the sovereignty of nation-states. As the Indian government's negative reaction to the appointment of an activist as a Commissioner illustrates, governments question the legitimacy of non-elected individuals as representatives of a broad view. Moreover, as governments' calls for no new conditionalities arising from the WCD show, they are wary of non-governmental actors' ability to circumscribe states' role through international agencies and such processes as the WCD.

What then is the pathway to implementation—one that captures the potential for creativity of multi-stakeholder processes, while recognising the legitimate role of governments? The full answer to this question must await the unfolding of reactions to the WCD report over time. However, the initial steps taken by various actors indicate a way forward.

The WCD Forum established a Dams and Development Unit (DDU) to carry forward its work. A range of Forum members—the World Bank, IUCN, an NGO, a river basin authority, a social movement, and a private sector actor—agreed to serve as the steering committee of this unit. The continued participation of this range of stakeholders demonstrates the ongoing relevance of the Commission's report.

Follow-up steps include efforts to reach out to governments. The WCD couched its recommendations within the context of the United Nations covenants and declarations on human rights, development, and environment. By so doing, it firmly located itself as within, rather than external to, the frameworks of intergovernmental deliberations. Thus, it provided a way for governments to engage with its findings in a manner that recognised the legitimacy of intergovernmental deliberations. Moreover, the steering committee's choice of an established intergovernmental body, the United Nations Environment Programme, as the host of the DDU, provides a further bridge to governments.

At the same time, rather than being backed by formal sanction mechanisms, widespread adoption of the WCD's recommendations depends on acceptance of norms of practice, supported by civil society scrutiny of the private sector, national governments, and international agencies. If successful, a critical role for the WCD will have been to crystallise and provide an impetus to norms of practice for infrastructure projects. Over the longer term, the bridge back to formal governmental and intergovernmental processes will likely be built incrementally, by incorporating practice into formal laws, in part through continued pressure by non-governmental actors.

This discussion reinforces the message that although democratisation of decision-making at the global level can bring significant advantages, ultimately advances in principles and practices must be translated to and implemented at the national level and below. However, as the experience of the WCD suggests, efforts at global and national democratisation are mutually reinforcing. In the WCD process, civil society organising at the national level served as the catalyst for creating the Commission and the seedbed for a transnational civil society alliance on dams. Conversely, the WCD process provided an avenue for greater expression at the national level and stimulated further dialogue across sectors at that level. The full potential of the World Commission on Dams—and other multi-stakeholder processes—lies in this promise of democratisation, at both the national and global levels.

Endnotes

1. Interviews with Forum members, September 2000 and November 2000. E-mail correspondence with Forum member, January 2001.
2. In interviews and public settings during the Forum meetings, government representatives expressed reservations about the extent of NGO and social movement participation in the WCD process. Also interview with government representative, April 2000.
3. Patrick McCully, "How to use a Trilateral Network: An Activist's Perspective on the World Commission on Dams." Paper presented at Agrarian Studies Program Colloquium, Yale University, 19 January 2001.
4. For example, the Dublin Principles agreed upon by governmental representatives in 1992 in the run-up to the United Nations Conference on Environment and Development recognised that "[the] pivotal role of women as providers and users of water and guardians of the living environment has seldom been reflected in institutional arrangements for the development and management of water" as one of four overarching principles. Principle Three of *The Dublin Statement*, International Conference on Water and the Environment: Development issues for the 21st century, 26-31 January 1992, Dublin, Ireland. The WCD's final report documents some of the effects of dam-related development and displacement on women, but its guidelines and recommendations incorporate only a passing mention of gender issues.
5. This recommendation is contained in the paper of an early advisor to the Commission, Anthony Dorcey, *Institutional Design and Operational Modalities for the Proposed Large Dams Commission*, Stockholm Draft, 6 August 1997 (mimeo).
6. Derived from Corporación Participa, Environmental Management and Law Association, Thailand Environment Institute, and World Resources Institute, "Framework for Assessing Public Access to Environmental Decision-Making," 2001.
7. Martha Finnemore and Kathryn Sikkink. Autumn 1998. "International Norm Dynamics and Political Change." *International Organisation* 52(4): 887-917.

Appendix 1

The World Commission on Dams and its Origins:
A Brief Chronology of Events

June 1994 .... Anti-dam organisations sign the Manibeli Declaration, calling for a moratorium on World Bank funding of large dams

September 1996 .... World Bank's Operations Evaluation Department (OED) Phase I review, *The World Bank's Experience With Large Dams: A Preliminary Review of Impacts*, released

March 1997 .... First International Meeting of People Affected by Dams, Curitiba Declaration signed

April 1997 .... International Rivers Network press release critiquing the OED review

April 1997 .... Gland, Switzerland meeting of World Bank, IUCN, and dam-related stakeholders

August 1997 .... Interim Working Group (IWG) meeting, Stockholm, Sweden

September 1997 .... Professor Kader Asmal chosen as WCD Chairperson

November 1997 .... WCD launch delayed

January 1998 .... "Expanded IWG" meeting in Cape Town, South Africa

April 1998 .... Achim Steiner chosen as WCD Secretary-General

May 1998 .... WCD established

May 1998 .... First meeting of the WCD in Washington, D.C.

September 1998 .... Jan Veltrop replaces Wolfgang Pircher as Commission's ICOLD representative

September 1998 .... India consultation cancelled

March 1999 .... First Forum meeting in Prague, Czech Republic

January 2000 .... Shen Guoyi formally resigns from the Commission

April 2000 .... Second Forum meeting in Cape Town, South Africa

November 2000 .... Release of *Dams and Development: A New Framework for Decision-Making*

February 2001 .... Third Forum meeting in Cape Town, South Africa

July 2001 .... WCD Secretariat officially closed, Dams and Development Unit established

Appendix 2

WCD Forum Members

Affected Peoples' Groups

Coordination for the Senegal River Basin (CODESEN), Senegal
Cordillera People's Alliance (CPA), Philippines
Federacíon de Indígenas del Estado Bolívar (COICA), Venezuela
Grand Council of the Cree (GCCEI), Canada
Movimento dos Antigos por Barragens (MAB), Brazil
Narmada Bachao Andolan (NBA), India
SUNGI Development Foundation, Pakistan

Bilateral Agencies / Export Credit Guarantee Agencies

Federal Ministry for Economic Cooperation and Development (BMZ),
Germany
Japan Bank for International Cooperation (JBIC), Japan
Norwegian Agency for International Cooperation (NORAD), Norway
Swedish International Development Agency (Sida), Sweden
Swiss Agency for Development and Cooperation (SDC), Switzerland
United States Export/Import Bank, United States

Government Agencies

Bureau of Reclamation, United States
Lesotho Highlands Water Project (LHWP), Lesotho
Ministry of Mahaweli Development, Sri Lanka
Ministry of Water Resources, China
Ministry of Water Resources, India
National Water Commission, Mexico

International Associations

International Association for Impact Assessments (IAIA)
International Commission for Irrigation and Drainage (ICID)
International Commission on Large Dams (ICOLD)
International Energy Agency (IEA)
International Hydropower Association (IHA)

Multilateral Agencies

African Development Bank (AfDB), Abidjan
Asian Development Bank (ADB), Manila
Inter-American Development Bank (IDB), Washington
United Nations Development Programme (UNDP), New York
United Nations Environment Programme (UNEP), Nairobi
United Nations Food and Agriculture Organization (FAO), Rome
World Bank (WB), Washington

NGOs

Berne Declaration, Switzerland
Development Alternatives with Women for a New Era (DAWN), Fiji
Environmental Development Action (ENDA), Senegal
Help the Volga River, Russia
International Rivers Network (IRN), United States
Intermediate Technology Development Group (ITDG), United Kingdom
Sobrevivencia-Friends of the Earth, Paraguay
Swedish Society for Nature Conservation (SSNC), Sweden
The World Conservation Union (IUCN), Switzerland
Transparency International (TI), Germany
Wetlands International, Japan
World Economy, Ecology & Development (WEED), Germany
World Wide Fund for Nature (WWF), Switzerland

Private Sector Firms

Asea Brown Boveri (ABB), Switzerland
Electric Power Development Corporation (EPDC), Japan
Enron, United States
Harza Engineering Firm, United States
Hopewell Holdings, Hong Kong
Saman Engineering Consultants, Korea
Siemens, Germany

Research Institutes/Resource Persons

Centro EULA, Ciudad Universitaria Concepcion, Chile
Focus on the Global South, Thailand
Institute of Hydroelectric Studies and Design (ISPH), Romania
International Water Management Institute (IWMI), Sri Lanka
Tropical Environmental Consultants, Ltd., Senegal
Water Research Institute (WRI), Israel Institute of Technology
Winrock International, Nepal
World Resources Institute (WRI), United States
Worldwatch Institute, United States
Wuppertal Institute, Germany

River Basin Authorities

Confederación Hidrográfica del Ebro (CHE), Spain
Jordan Valley Authority (JVA), Jordan
Mekong River Commission (MRC), Cambodia
Volta River Authority (VRA), Ghana

Utilities

Electricité de France, France
Electrobras, Brazil
Hydro-Québec, Canada
Mini Hydro Division (MHD), Philippines
Nepal Electricity Authority (NEA), Nepal

Source: World Commission on Dams, Dams and Development: A New Framework for Decision-Making (London: Earthscan, 2000).

Appendix 3
Thematic Reviews

Social

I.1 Social impacts of large dams equity and distributional issues

I.2 Dams, indigenous people, and vulnerable ethnic minorities

I.3 Displacement, resettlement, rehabilitation, reparation, and development

Environment

II.1 Dams, ecosystem functions, and environmental restoration

II.2 Dams and global change

Economic

III.1 Economic, financial, and distributional analysis

III.2 International trends in project financing

Options Assessment

IV.1 Assessment of electricity supply and demand management options

IV.2 Assessment of irrigation options

IV.3 Assessment of water supply options

IV.4 Assessment of flood control and management options

IV.5 Operation, monitoring, and decommissioning of dams

Institutional Processes

V.1 Planning approaches

V.2 Environmental and social assessment for large dams

V.3 River basins-institutional frameworks and management options

V.4 Regulation, compliance, and implementation options

V.5 Consultation and decision-making processes

Source: World Commission on Dams, Dams and Development: A New Framework for Decision-Making (London: Earthscan, 2000).

Appendix 4

Strategic Priorities of the Dams and Development *Report*

Strategic Priority 1 – Gaining Public Acceptance

1.1 Recognition of rights and assessment of risks are the basis for the identification and inclusion of stakeholders in decision-making on energy and water resources development.

1.2 Access to information, legal and other support is available to all stakeholders, particularly indigenous and tribal peoples, women and other vulnerable groups, to enable their informed participation in decision-making processes.

1.3 Demonstrable public acceptance of all key decisions is achieved through agreements negotiated in an open and transparent process conducted in good faith and with the informed participation of all stakeholders.

1.4 Decisions on projects affecting indigenous and tribal peoples are guided by their free, prior and informed consent achieved through formal and informal representative bodies.

Strategic Priority 2 – Comprehensive Options Assessment

1.1 Development needs and objectives are clearly formulated through an open and participatory process before the identification and assessment of options for water and energy resource development.

1.2 Planning approaches that take into account the full range of development objectives are used to assess all policy, institutional, management, and technical options before the decision is made to proceed with any programme or project.

1.3 Social and environmental aspects are given the same significance as technical, economic and financial factors in assessing options.

1.4 Increasing the effectiveness and sustainability of existing water, irrigation, and energy systems are given priority in the options assessment process.

1.5 If a dam is selected through such a comprehensive options assessment process, social and environmental principles are applied in the review and selection of options throughout the detailed planning, design, construction, and operation phases.

Strategic Priority 3 – Addressing Existing Dams

1.1 A comprehensive post-project monitoring and evaluation process, and a system of longer-term periodic reviews of the performance, benefits, and impacts for all existing large dams are introduced.

1.2 Programmes to restore, improve and optimise benefits from existing large dams are identified and implemented. Options to consider include rehabilitate, modernise and upgrade equipment and facilities, optimise reservoir operations and introduce non-structural measures to improve the efficiency of delivery and use of services.

1.3 Outstanding social issues associated with existing large dams are identified and assessed; processes and mechanisms are developed with affected communities to remedy them.

1.4 The effectiveness of existing environmental mitigation measures is assessed and unanticipated impacts identified; opportunities for mitigation, restoration and enhancement are recognised, identified and acted on.

1.5 All large dams have formalised operating agreements with time-bound licence periods; where replanning or relicensing processes indicate that major physical changes to facilities or decommissioning may be advantageous, a full feasibility study and environmental and social impact assessment is undertaken.

Strategic Priority 4 – Sustaining Rivers and Livelihoods

1.1 A basin-wide understanding of the ecosystem's functions, values and requirements, and how community livelihoods depend on and influence them, is required before decisions on development options are made.

1.2 Decisions value ecosystems, social and health issues as an integral part of project and river basin development and prioritise avoidance of impacts in accordance with a precautionary approach.

1.3 A national policy is developed for maintaining selected rivers with high ecosystem functions and values in their natural state. When reviewing alternative locations for dams on undeveloped rivers, priority is given to locations on tributaries.

1.4 Project options are selected that avoid significant impacts on threatened and endangered species. When impacts cannot be avoided viable compensation measures are put in place that will result in a net gain for the species within the region.

1.5 Large dams provide for releasing environmental flows to help maintain downstream ecosystem integrity and community livelihoods and are designed, modified and operated accordingly.

Strategic Priority 5 – Recognising Entitlements and Sharing Benefits

1.1 Recognition of rights and assessment of risks is the basis for identification and inclusion of adversely affected stakeholders in joint negotiations on mitigation, resettlement and development related decision-making.

1.2 Impact assessment includes all people in the reservoir, upstream, downstream and in catchment areas whose properties, livelihoods and non-material resources are affected. It also includes those affected by dam related infrastructure such as canals, transmission lines and resettlement developments.

1.3 All recognised adversely affected people negotiate mutually agreed, formal and legally enforceable mitigation, resettlement and development entitlements.

1.4 Adversely affected people are recognised as first among the beneficiaries of the project. Mutually agreed and legally protected benefit sharing mechanisms are negotiated to ensure implementation.

Strategic Priority 6 – Ensuring Compliance

1.1 A clear, consistent and common set of criteria and guidelines to ensure compliance is adopted by sponsoring, contracting and financing institutions and compliance is subject to independent and transparent review.

1.2 A Compliance Plan is prepared for each project prior to commencement, spelling out how compliance will be achieved with relevant criteria and guidelines and specifying binding arrangements for project-specific technical, social and environmental commitments.

1.3 Costs for establishing compliance mechanisms and related institutional capacity, and their effective application, are built into the project budget.

1.4 Corrupt practices are avoided through enforcement of legislation, voluntary integrity pacts, debarment and other instruments.

1.5 Incentives that reward project proponents for abiding by criteria and guidelines are developed by public and private financial institutions.

Strategic Priority 7 – Sharing Rivers for Peace, Development and Security

1.1 National water policies make specific provision for basin agreements in shared river basins. Agreements are negotiated on the basis of good faith among riparian States. They are based on principles of equitable and reasonable utilisation, no significant harm, prior information and the Commission's strategic priorities.

1.2 Riparian States go beyond looking at water as a finite commodity to be divided and embrace an approach that equitably allocates not the water, but the benefits that can be derived from it. Where appropriate, negotiations include benefits outside the river basin and other sectors of mutual interest.

1.3 Dams on shared rivers are not built in cases where riparian States raise an objection that is upheld by an independent panel. Intractable disputes between countries are resolved through various means of dispute resolution including, in the last instance, the International Court of Justice.

1.4 For the development of projects on rivers shared between political units within countries, the necessary legislative provision is made at national and sub-national levels to embody the Commission's strategic priorities of 'gaining public acceptance', 'recognising entitlements' and 'sustaining rivers and livelihoods'.

1.5 Where a government agency plans or facilitates the construction of a dam on a shared river in contravention of the principle of good faith negotiations between riparians, external financing bodies withdraw their support for projects and programmes promoted by that agency.

Appendix 5

Events Attended by the Assessment Team

WCD Events	Date & Attendee(s)
Latin America consultation, Brazil	12-13 August 1999 Manuel Pulgar-Vidal,[a] WRI
WCD presentation to World Bank, Washington	28 September 1999 Mairi Dupar, WRI
Africa and Middle East consultation, Egypt	8-9 December 1999 Mairi Dupar, WRI
USA stakeholder meeting	13 January 2000 Tundu Lissu, LEAT
European NGO meeting, Slovakia	17-18 January 2000 Elena Petkova, WRI
Pakistan stakeholder meeting	17-18 January 2000 Gopal Siwakoti "Chintan,"[b] Lokayan
Turkey stakeholder meeting	20-21 January 2000 Elena Petkova, WRI
Norway stakeholder meeting	14 February 2000 Smitu Kothari, Lokayan
Zambia stakeholder meeting	21-22 February 2000 Melchesideck Lutema , LEAT
India stakeholder meeting	21, 23 February 2000 Smitu Kothari, Lokayan Anil Bhattarai , Lokayan
East and Southeast Asia consultation, Vietnam	26-27 February 2000 Mairi Dupar, WRI Anil Bhattarai , Lokayan Melchesideck Lutema , LEAT
WCD second Forum meeting, South Africa	6-8 April 2000 Mairi Dupar, WRI Navroz Dubash, WRI Smitu Kothari, Lokayan Anil Bhattarai, Lokayan Gopal Siwakoti "Chintan,"[b] Lokayan Tundu Lissu, LEAT
WCD third Forum meeting, South Africa	25-27 February 2001 Mairi Dupar, WRI Navroz Dubash, WRI

Events Related to Dams **(not part of official WCD process)**	Date & Attendee(s)
Discussions with stakeholders, Thailand	January 2000 Gopal Siwakoti "Chintan,"[b] Lokayan
World Water Forum, Netherlands	March 2000 Gopal Siwakoti "Chintan,"[b] Lokayan
Meeting of dam-related stakeholders, Nepal	June 2000 Anil Bhattarai, Lokayan Gopal Siwakoti "Chintan,"[b] Lokayan
Conference on Mekong regional environmental governance, Laos	April 2001 Mairi Dupar, WRI
Meeting of dam-related stakeholders, India	May 2001 Smitu Kothari, Lokayan Ramananda Wangkheirakpam,[c] Lokayan Lakshmi Rao,[d] Lokayan

[a] Representing WRI, primary affiliation with the Peruvian Society of Environmental Law
[b] Representing Lokayan, primary affiliation with INHURED International
[c] Representing Lokayan, primary affiliation with Jawaharlal Nehru University
[d] Representing Lokayan, primary affiliation with Jawaharlal Nehru University

Who We Are

World Resources Institute

World Resources Institute provides—and helps other institutions provide—information and practical proposals for policy and institutional change that will foster environmentally sound, socially equitable development. WRI researches and publicises policy options, encourages adoption of innovative approaches, and provides strong technical support to governments, corporations, international institutions, and environmental non-governmental organisations (NGOs). WRI's Institutions and Governance Program, a program that focuses on the social and political dimensions of environmental policymaking, houses the WCD assessment team.

World Resources Institute, 10 G Street, NE; Suite 800; Washington, DC 20002, USA

Lokayan

Lokayan, a 20-year-old action-research centre in India, works with social movements, research institutes, policy-makers and citizens at large to foster the widening of justice, democracy, and ecological sustainability. It does this through participatory research, campaigns, advocacy, political mobilisation, dialogues, and publications. Lokayan won the Right Livelihood Award ("Alternative Nobel Prize") in 1986.

Lokayan, 13 Alipur Road, Exchange Building, Civil Lines, Delhi 110 054, India

The Lawyers' Environmental Action Team

The Lawyers' Environmental Action Team (LEAT) is a public interest lawyers' organisation based in Dar es Salaam, Tanzania. Founded in 1994, LEAT specialises in public policy research and advocacy in the field of environment and natural resource management. It has undertaken applied policy research work on institutional and governance issues for government departments and donor agencies. It also carries out public interest litigation on selected cases on behalf of rural communities.

LEAT, Kings Palace Hotel Building, First Floor, Sikukuu Street, Kariakoo Area, P.O. Box 12605, Dar Es Salaam, Tanzania